Birds of Lesotho

A guide to distribution
past and present

Kurt Bonde

UNIVERSITY OF NATAL PRESS
PIETERMARITZBURG
1993

© University of Natal 1993
P.O. Box 375
Pietermaritzburg 3200
South Africa

All rights reserved. No part of this publication, including illustrations, photographs and maps, may be reproduced or transmitted, in any form or by any means, without permission of the copyright holders.

Cover photograph and frontispiece © Nigel Dennis/ABPL

ISBN 0 86980 881 8

Cover Cape Vulture (*Gyps coprotheres*)
Frontispiece Bearded Vulture (*Gypaetus barbatus*)

Typeset in the University of Natal Press
Printed by Kohler Carton and Print
P.O. Box 955
Pinetown 3600
South Africa

To all those who, through their fieldwork, made this book possible. And to those who, through their fieldwork, will make this work out of date.

Above *A summer scene with Sotho horsemen on top of the Drakensberg, in the Afromontane zone.* (Photograph: G. Symons)

Below *The Sani River, a clear mountain stream in degraded Afromontane habitat with stony substrate: typical birds include Mountain Pipit, Thickbilled Lark, Stonechat and Sentinel Rock Thrush.* (Photograph: R. Guy)

Contents

Acknowledgements viii

Foreword ix

Preface xi

Lesotho landscapes 1

Ornithological history 4

Birding in Lesotho 11

Gazetteer 23

References
   Birds of Lesotho 29

General references 35

Notes on the checklist 38

Annotated checklist 39

Index to English names 101

Index to scientific names 104

Acknowledgements

The author and publishers would like to thank the following individuals and institutions for their assistance and permission to reproduce illustrations.

Tony Clarkson for his bird drawings, the Anthony Bannister Photo Library for the cover picture and frontispiece, Robin Guy and Godfrey Symons for scenic photographs.

Helena Margeot of the University of Natal, Pietermaritzburg, Cartographic Unit for redrawing the author's sketch maps.

David Moon for the cover design.

A special word of thanks is due to Gordon Maclean for providing the Foreword, for his enthusiasm and for cheerfully tendered and valuable advice.

Foreword

Nearly 30 years ago the Percy FitzPatrick Institute of African Ornithology produced a booklet by Charlot Jacot-Guillarmod entitled *Catalogue of the birds of Basutoland*. This has been hitherto the only work on the avifauna of the small Protectorate. Since then much has changed. The name of the country is now Lesotho, its human population has grown tremendously and its environments are under increasing threats of degradation from overexploitation of many kinds. The appearance of Kurt Bonde's *Birds of Lesotho* is therefore timeous. Not only will this book create awareness of birds among the Basotho, but it will give fresh impetus to an interest in the birds of Lesotho by travellers to that country. Let us hope that, in the longer term, these effects will lead to more widespread and enlightened conservation measures which in turn will result in a better quality of life for the avian and human inhabitants of this beautiful land. That the material was collected several years ago, and that southern Africa is to be furnished with a comprehensive bird atlas in a few years' time are unimportant against the background of the long time lapse since the publication of Jacot-Guillarmod's booklet and the relatively poor coverage that Lesotho will have received from atlassers, even in more recent years. The University of Natal Press has acted with commendable foresight in taking on this small but significant work. If the environmental conservation of Lesotho becomes a reality, this book will have served a major purpose. If, however, Lesotho loses much of its birdlife through careless land use, this book will be a memorial to what once was.

Gordon Lindsay Maclean

Above *Overgrazed Afromontane meadow at the top of Sani Pass: typical birds include White Stork (in summer), Black Crow, Bald Ibis, Redcapped Lark and Sicklewinged Chat.* (Photograph: R. Guy)

Below *A winter scene after a snowfall in the Afromontane zone.* (Photograph: G. Symons)

Preface

Having arrived in Lesotho in March 1978 and being interested in birds I looked for some information on their current status. I quickly realized that there was no comprehensive work available and so the idea arose of arranging my own field notes into an article. During 1980 this 'article' grew into a preliminary checklist in which all my collected information on each species, both historical and current, was included. I hoped this would form the basis from which other birdwatchers could work, adding their own findings and enabling any change in status of species to be recorded. As a rule I have maintained a neutral approach to the validity of records, especially as many are old and it is impossible to corroborate their findings.

Before leaving Lesotho in April 1981 this preliminary checklist (mimeographed in a small number) was released and distributed to various persons and institutions in Lesotho, South Africa, the U.K. and the U.S.A. In this way I hoped to collect more unpublished material from birdwatchers and to elicit comments and criticism. I have also continued to search for written records and have contacted museums to inquire whether skins or eggs from Lesotho were included in their collections. The information thus obtained has now been included in this edition.

The main body of this book is the Checklist. For the reader interested in the broader aspects of Lesotho's natural history may I recommend the splendid *Guide to Lesotho* by David Ambrose.

Originally I had planned that this edition should be arranged in the same way as the preliminary checklist of 1981, where all collected material was quoted without change. This would, however, have made the book too bulky and unwieldy, so the text has been rearranged. A copy of the final draft has been placed in the library of the Percy FitzPatrick Institute of African Ornithology, University of Cape Town.

The assistance of the following institutions is acknowledged: The British Museum, London; Edward Grey Institute of Field Ornithology, London; Transvaal Museum, Pretoria; Durban Museum; East London Museum; Port Elizabeth Museum; Percy FitzPatrick Institute of African Ornithology, Cape Town; National Museums and Monuments of Zimbabwe, Bulawayo; Bioscience Information Service and Zoological Record, London. I also wish to thank the following persons: H.G.M. Bass, David Boddam-Whetham, Neville Brickell, Richard Brooke, Christopher Brown, Victor Burke, Clive Clements, Dave Coghlan, Digby Cyrus, Donald Davidson, Reginald Dove, Tibor Farkas, David Halsted, John Jilbert, P.J. Jones, R. Liversidge, Harry Loots, Geoff McLachlan, Manfred Reichardt, Nigel Robson, Gustav Rudebeck, Richard Ryan, L.G.A. Smits, Walter Stanford, Philip Steele, Godfrey Symons, Jack Vincent and John Williamson. I am especially grateful to David Ambrose, who from the start gave me the inspiration for this book, and to Pauline Jilbert, who at a critical moment stepped in and helped to edit the introductory chapters.

Kurt Bonde

Lesotho landscapes

The Kingdom of Lesotho is a small country extending over 30 350 km². It lies between the latitudes of 28° and 31° south and longitudes of 27° and 30° east. About a quarter of the country in the west can be classified as lowlands varying in height above sea-level from about 1500 metres to 1800 metres; the remaining three-quarters are highlands rising to a height of over 3500 metres in the Drakensberg range which forms the eastern boundary with the Natal province of South Africa.

Lesotho has traditionally been divided into these two major parts, but this is not a true picture of the country. Phrases like 'the Lesotho massif' are used, which reinforce the impression of a high plateau as opposed to the lowlands. This is too simplistic a division. The mountains are in fact traversed by rivers, all of which have their sources high in the Drakensberg where the border between Lesotho and South Africa follows the escarpment. These rivers have, through erosion, attained maturity not far from their sources deep in the mountains. The valleys of the Senqu and its tributaries form corridors of lowland-like environments deep into the mountain massif, creating steep gradations in relief, altitude and habitat within short distances of the rivers themselves. Several 'lowland' species are actually found in the heart of the mountains.

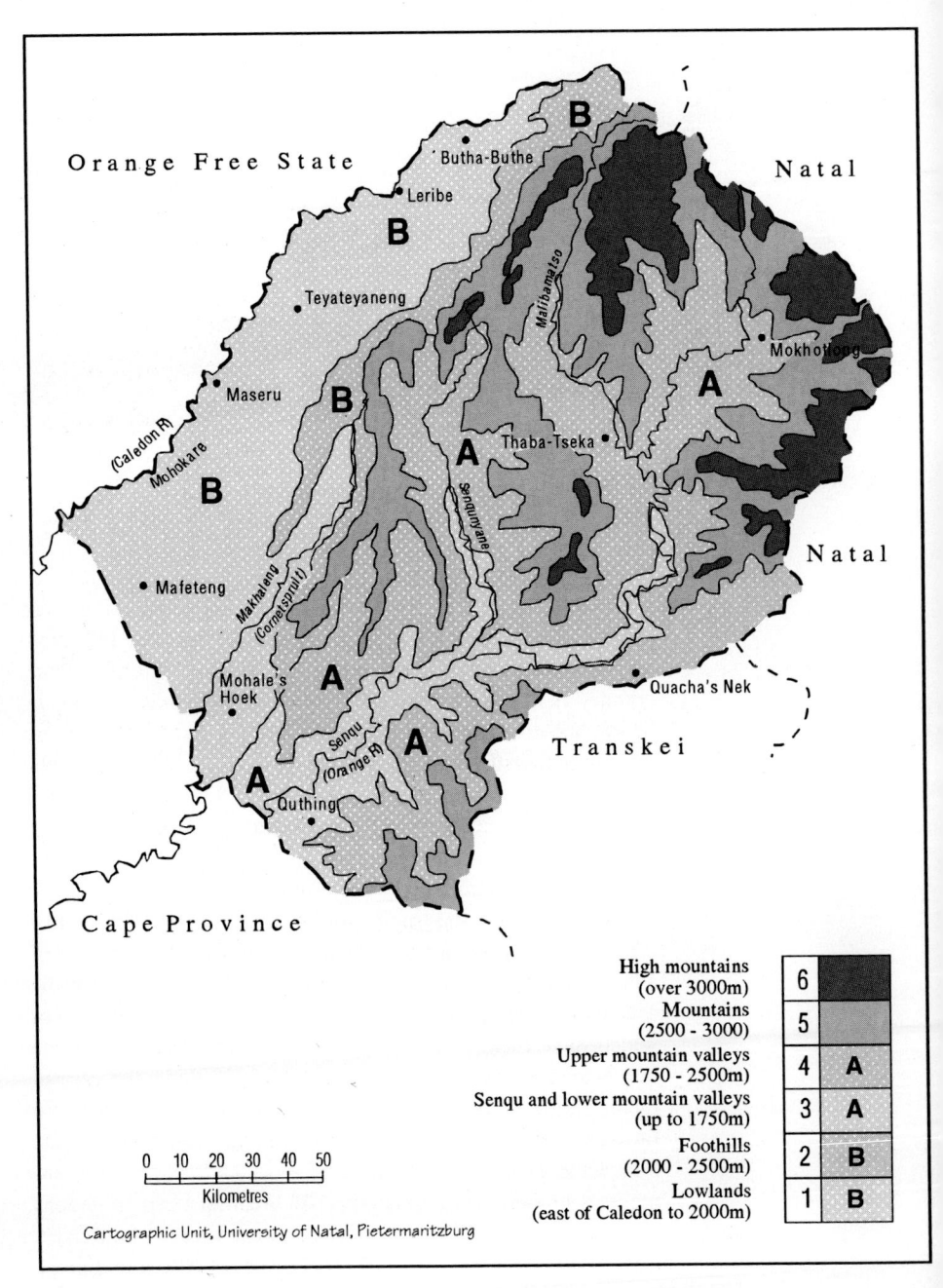

Figure 1

From an ecological point of view it is important to define the regional structure in greater detail. Lesotho may be divided into six regions (see figure 1):

1. **Lowlands** From the Caledon east to 2000 metres (all heights above sea-level); 21,25% of the country.
2. **Foothills** Between 2000 metres and 2500 metres, bordering the lowlands; 11,76% of the country.
3. **Senqu Valley and lower mountain valleys** The upper limit of this region is 1750 metres; it comprises 21,83% of the country. Definition between this region and the lowlands becomes difficult at the southern end of the Senqu Valley. The Maphutseng River has been taken as the dividing line.
4. **Upper mountain valleys** Between 1750 metres and 2500 metres; 15,75% of the country.
5. **Mountain region** Between 2500 metres and 3000 metres; 12,86% of the country.
6. **High mountain region** Above 3000 metres; 16,55% of the country.

This regional division is a modification of an original classification made by L.G.A. Smits (*in litt.*).

These concepts have to some extent been worked into the Checklist; information covering regions 3 to 6 is somewhat thin though and it is best to avoid making categorical statements about the distribution of species within these areas. I hope, however, that this regional division will prove useful and bring about a better understanding of the distribution of birds in Lesotho.

Ornithological history

Shortly after the foundation of the Basotho Nation by King Moshoeshoe who settled at Thaba-Bosiu around 1824, the lowlands of present-day Lesotho were visited by the 'Expedition for Exploring Central Africa' led by Dr, later Sir Andrew Smith. This expedition entered Lesotho near Likhoele Mountain on 12 October 1834, and visited Morija and Thaba Bosiu. On 27 October it left Lesotho, crossing the Caledon a little north of present-day Maseru. On 8 November the expedition again crossed the Caledon north of Ficksburg, turned south, and after crossing the Hlotse River, went further south to the Phuthiatsana River near Teyateyaneng. This river was followed eastward, and a party of the expedition ascended the first mountain range on 18 November. The expedition thereafter turned west again and left Lesotho on 24 November 1834.

In 1836, the *Report of the Expedition for Exploring Central Africa* . . . was published by the Government Gazette Office, Cape Town. It is today a very rare piece of Africana. An abridged version was published in the *Journal of the Royal Geographical Society of London*, (1836), and the *Illustrations of the Zoology of South Africa*, (1838–1849). The other papers from this expedition were published only much later: Andrew Smith's diary in 1939–40, the journal of the expedition in 1975 and the diary of one of the participants, John Burrow, in 1971. In this material a few bird sightings are mentioned. As Andrew Smith's writings have only recently been published this expedition contributed little to the study of Lesotho's birdlife, and later expeditions simply bypassed Lesotho on their way north.

Layard's *Birds of South Africa* (1875–84) makes no mention of Lesotho. The *History of the Collections contained in the Natural History Departments of the British Museum*, vol. 2, (1906), reveals that neither eggs nor skins had been received from Lesotho.

The next major book on the birds of the subcontinent, Stark & Sclater's *Birds of South Africa* (1901–06), refers a few times to Basutoland, but contains nothing substantial. The information was obtained from J.P. Murray, a civil servant in Lesotho who resided in Mafeteng and Maseru from approximately 1884 to 1926. Apart from a few notes in journals, he did not publish anything. In 1964 the Percy FitzPatrick Institute published some of the notes he had made in his copies of 'Layard' and 'Stark & Sclater' in their *South African Avifauna Series* (no. 21).

The first substantial article on Lesotho's avifauna was written by Roden E. Symons, a game warden at Giant's Castle Game Reserve in Natal from 1906 to 1916. This is of special interest as it deals with birds from the mountain and high mountain regions. Thirty years later Colonel Jack Vincent collected a number of skins from the same area. These are now with the British Museum, being the only skins from Lesotho in that collection.

In the 1920s, J.A. Cottrell collected eggs while visiting Lesotho. This collection, which seems to be the only one from Lesotho, is now with the National Museum in Bulawayo, Zimbabwe.

The next article was by Charlot Jacot-Guillarmod (1932) who lived most of his life in Lesotho. In this article he notes the relative poverty of birdlife in Lesotho, an idea repeated elsewhere. He later wrote the most important document to date on the birdlife of the country. His *Catalogue of the birds of Basutoland* appeared in 1963. In the 1930s he collected many birds which were sent to the Transvaal Museum and almost all the skins in this collection were contributed by him.

In the 1950s and 1960s a number of scientific expeditions included Lesotho in their field of interest. In 1950–51 a party from Lund University in Sweden, during a tour of southern Africa, visited briefly. In 1956 the Durban Museum made a collecting trip, taking 155 skins mainly along the Mountain Road around Molimo Nthuse. In 1967 and 1968 the East London Museum collected 302 specimens, mainly on points east of Blue Mountain Pass. C.D. Quickelberge (1972), writing up this expedition, agreed with Jacot-Guillarmod that birds were scarce in the populated areas, but commented that they were easily seen in the mountains.

Even though Jacot-Guillarmod's book is a milestone, only one short notice was published in response, by C.F. Goodfellow (1966). He was able to add a few more birds to the *Catalogue*. He was a lecturer at the National University in Roma. After his death his ornithological notes were given to the Percy Fitz-Patrick Institute in Cape Town. Most of them deal with the Cape Bunting, its biology and behaviour.

Since then two more interested individuals have kept records; K.N.C. MacLeay and Angela Aspinwall. Both published reports of their sightings in Roma Valley, but as these are mimeographed it is doubtful that they have been known to a broader audience. In 1973 the Lesotho Ornithological Club was formed on a rather informal basis, as no officers were ever elected. The club brought out, as far as I know, four issues of a periodical named *Linonyana tsa Lesotho* (Birds of Lesotho). The initiator was Berton M. Bailey, then director of the United States Information Service in Maseru. I have been able to trace only the first two issues. Resident biologists in Sehlabathebe National Park have kept a record of the sightings in the park.

Victor Burke, based in Maseru between 1972 and 1978, intended to write a book on the birds of Lesotho, but never got around to writing it and he has kindly forwarded to me his own notes together with the notes he had collected, and these I have attempted to incorporate in the Checklist.

In 1980, Manfred Reichard's book *Tourist guide to the birds of Lesotho* was published. This contains many excellent photographs of birds of the region, but it also contains serious errors; for example, the Fish Eagle has not been recorded in Lesotho, and the single record of the Greater Kestrel is old and dubious. I have not quoted this book as a source in the Checklist as it is not a record of actual sightings.

In 1981 my preliminary checklist was released and gave rise to a large correspondence with persons and institutions. This has been the source of further material which has been incorporated into this volume.

In its efforts to protect the flora and fauna of Lesotho, the Protection and Preservation Commission (P.P.C.) has launched a project to map the occurrence of plant and animal life in Lesotho.

Even though it is 150 years since Sir Andrew Smith visited Lesotho, written records of the country's birdlife remain few and rather sketchy. It is my hope that this publication will bring together what information there is and will stimulate others to expand that information so that the diversity of Lesotho's birdlife can be appreciated by an ever-increasing number of bird enthusiasts.

Coverage

My summary of the current state of recorded knowledge is illustrated in figure 2. The lowland region (3a) is generally well covered though not every locality has been covered. The Maseru/Roma area (4), however, has been the home of numerous birdwatchers who have kept field notes, so the knowledge of birdlife in that part of the lowlands is particularly good.

The area 3b is that covered by the two collecting expeditions along the Mountain Road and, although these were for a brief period only, they provide valuable information on the foothill, upper mountain valley and mountain regions.

More information on the upper mountain valley, mountain and high mountain regions comes from the north-east corner (3c), Mokhotlong district, where Roden Symons (1916), Jack Vincent (*in litt.*), Godfrey Symons (*in litt.*) and Christopher Brown (*in litt.*) have kept records.

Sehlabathebe National Park (3d) covers the upper mountain valley and mountain regions. The monthly records kept there, however, do not make this regional distinction in locating the sightings. Some useful information can be gleaned from these records, nonetheless. More mountain region birds have been recorded from the north (2) by Jacot-Guillarmod (1963) from notes supplied by G. Maclean.

Thus we are left with a large area (1) where no systematic recording has taken place. In the Checklist you will find a few records here and there, but the gaps are enormous; nothing from

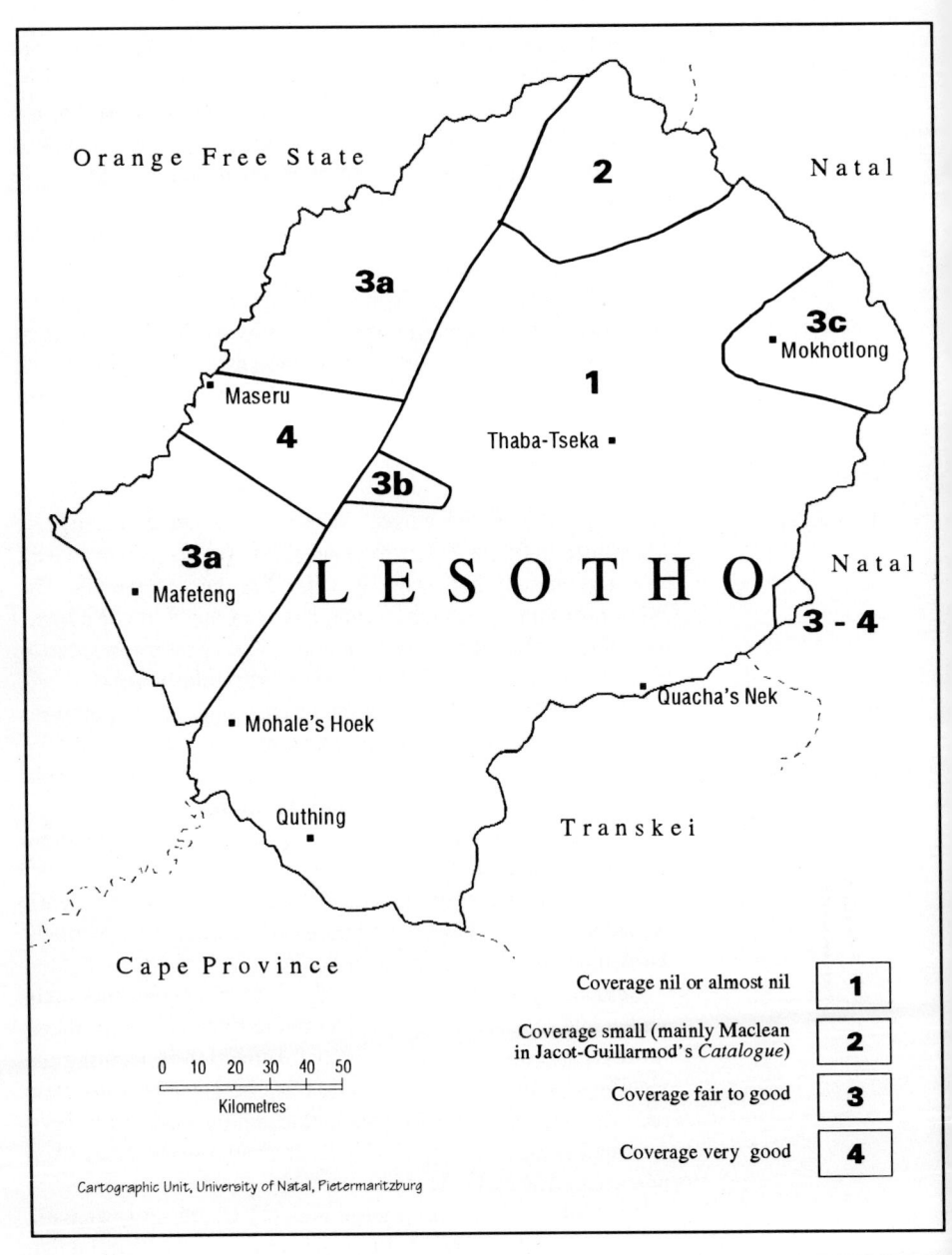

Figure 2

Thaba-Tseka and hardly anything from Qacha's Nek district apart from Sehlabathebe. Quthing district is almost untouched, apart from sightings from a few selected spots such as Koma-Koma Bridge and the Mphaki area (both upper mountain valley region), and Ha Sempe which is in the lower mountain valley region.

It must, however, be appreciated that this map reflects the road situation and the difficulties involved in travelling outside the lowlands, especially in former years. Now the road to Thaba-Tseka should be passable for the normal family car with only two-wheel-drive, and plans are afoot to link Thaba-Tseka and Mokhotlong with a tarred road. The Southern Perimeter Road is intended to link the lowlands with Qacha's Nek, so in the coming years it should be possible to reach most of the country in a two-wheel-drive vehicle. This should open up the country to the average birdwatcher.

It can, therefore, be hoped that the highlands in the coming years will be visited by many more birdwatchers, so more information can be collected, and a clearer picture of Lesotho's birdlife may emerge.

Birding in Lesotho

This chapter gives a few indications of good spots to see birds. I have not mentioned places like Semongkong and Sehlabathebe that are noted in more general tourist guides. It is intended to offer a starting point to anyone new to birdwatching in Lesotho.

Maseru area

In and around Maseru there are some easily accessible areas suitable for an afternoon stroll, just to keep the dust from the binoculars (see figure 1).

Along the Mohokare/Caledon

I visited two stretches along the river: below the polo ground and downstream to the Holiday Inn. I did not continue further downstream as this meant approaching the PMU (police) Barracks. It is never wise to get too close to military areas with anything that looks like a camera. The other area visited stretches from the sewage ponds, upstream past the water works purification plant and Lerotholi Technical School. I did not go downstream, as this meant coming close to the Maseru Bridge border post. On both stretches it is fairly easy to follow the river next to the water when the level is not too high. In the trees and scrub along the river, I have seen Cardinal Woodpecker, Lesser Honeyguide and Cape White-eye.

The golf course

If you observe some simple rules (watch out for flying balls as well as birds, keep off the greens, don't obstruct play) the golf course can provide some pleasant birdwatching. In the tall trees and dense scrub following the course of a small stream, I have seen Redfaced Mousebird, Olive Thrush and Redeyed Bulbul. Near the south fence the stream flows into a dam where I have heard African Marsh Warbler and seen the Cape Weaver building its nest in the reeds.

Maseru West and golf course/Orpen Road area

The long-established gardens of this area with their old trees and thick hedges attract many species. I have seen Laughing Dove, Cape Robin, Bokmakierie, three species of sparrow and breeding African Hoopoes, and in season Lesser Kestrel together with a few Eastern Redfooted Kestrel in the tall eucalyptus trees near the King's Palace.

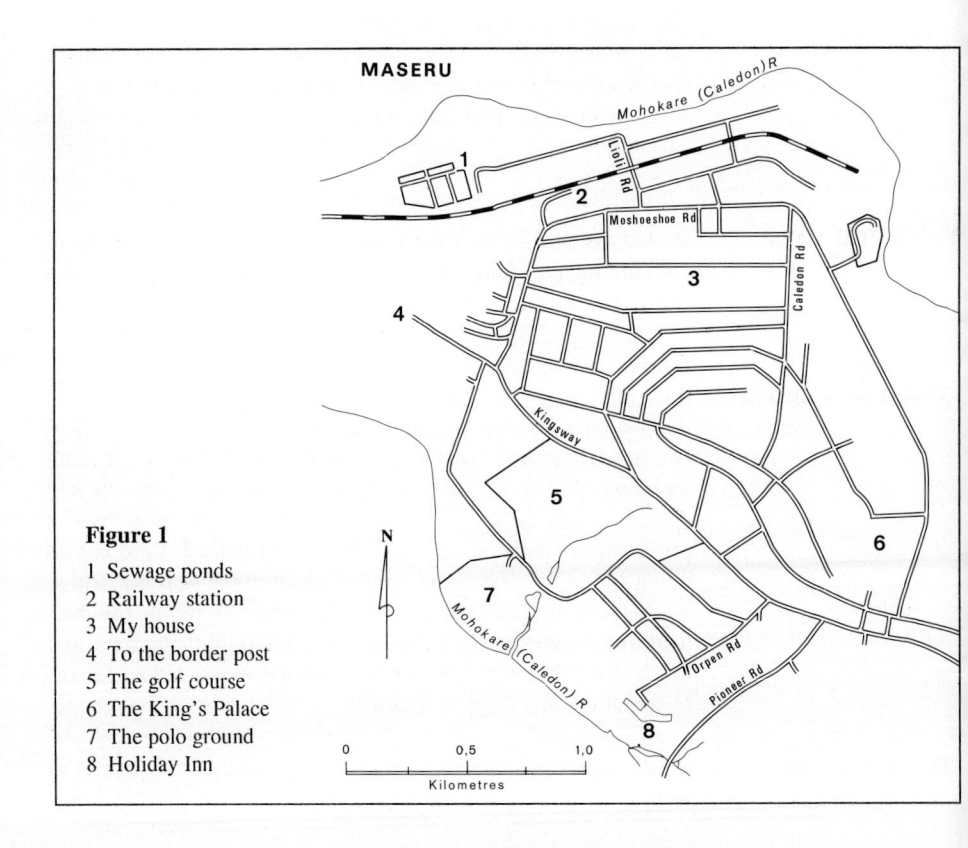

Figure 1
1 Sewage ponds
2 Railway station
3 My house
4 To the border post
5 The golf course
6 The King's Palace
7 The polo ground
8 Holiday Inn

Sewage ponds

To reach these, turn off from Moshoeshoe Road (sometimes known as Industrial Road) into Lioli Road and follow this to the end. On this route I have seen Glossy Starling and Redheaded Finch. The ponds usually have a strong smell, but depending on the water level they are worth visiting, as many ducks and waders can be seen as well as Giant Kingfisher, swifts and swallows.

For anyone staying in Maseru West, a walk through the industrial area, down to the sewage ponds and then along the Mohokare past the water works, can be strongly recommended.

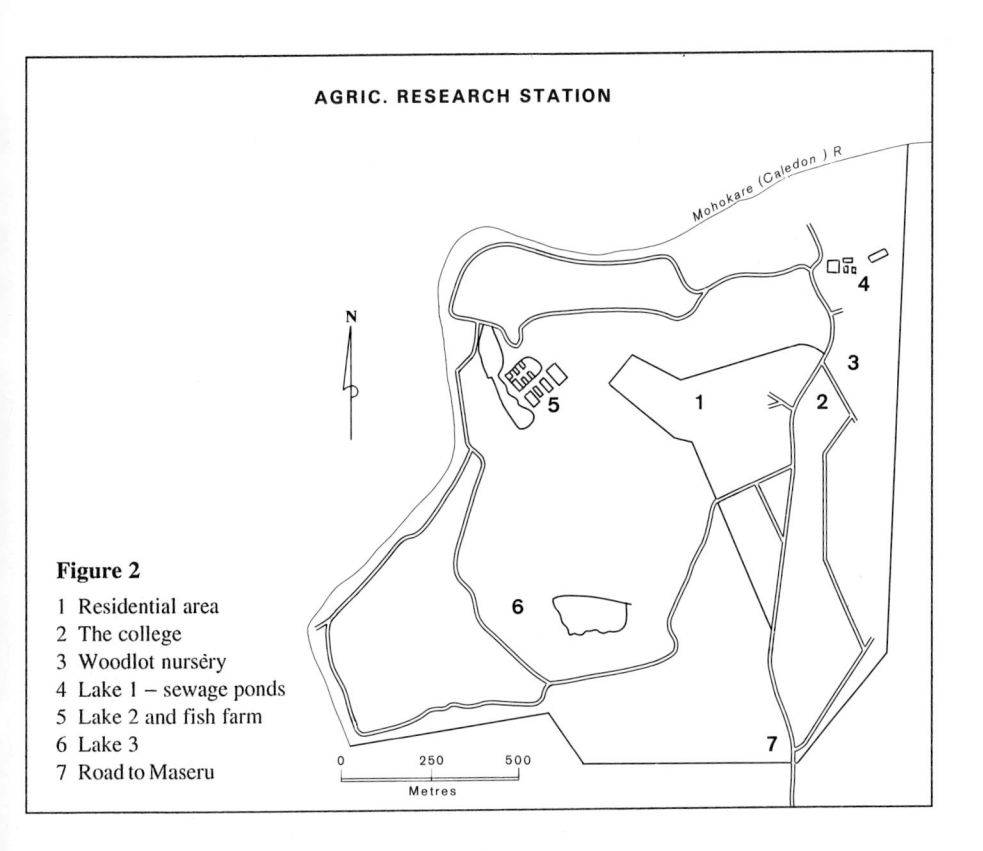

Figure 2
1 Residential area
2 The college
3 Woodlot nursèry
4 Lake 1 – sewage ponds
5 Lake 2 and fish farm
6 Lake 3
7 Road to Maseru

Agricultural Research Station and Lesotho Agricultural College

To reach this very fine birding locality, take the North Road from the Circle, turn left onto Rantsala Road, pass the stadium and the terminal building of the old airport and then turn right along Airport Road, or alternatively take Moshoeshoe Road from the Circle and turn right at the first robot (Airport Road), travel 750 metres through the suburb of Thibella until the road turns sharply left. Follow this road to the end and you will be at the Agricultural Research Station. This is the best area in Maseru for seeing a wide variety of birds. It is possible to make a circular drive of the Station (see figure 2) though at times there are locked gates and it is necessary to walk. Along the route you will find some water stretches: sewage ponds (figure 2, lake 1), a dam and fish farm ponds (lake 2) and another more open dam (lake 3). The fish farm area has provided sightings of Darter, Little Egret, Little Bittern, Blackcrowned Night Heron, Purple Gallinule and many passerine birds. Lake 3 often has ducks and waders, together with cormorants and, in the reeds, Cape Reed Warbler, African Marsh Warbler, and also Cape Robin. In the fields Helmeted Guineafowl, Swainson's Francolin, Quail Finch and Desert Cisticola can be seen. I have observed Cape Robin, Olive Thrush and Cape White-eye in the trees by the river, and Black Duck on the river.

North from Maseru

If one takes the North Road out of Maseru there are a number of interesting sites en route: the Marabeng Dam, Leshoboro Woodlot, Nye-Nye Dam and just north of Leribe, the Leribe Dam.

Marabeng Dam

This is the first dam you encounter, about 10 km out of Maseru, at a point where the road turns sharply left. I stopped there once (4 February 1979) and saw only a few birds, of which only Redeyed Dove is noteworthy. I have learned since that Cattle Egret bred in the trees around the dam from 1973 to 1975. They were also thought to have bred at Tebetebeng Mill on the Mamathe's road north of Teyateyaneng, so both these locations need to be checked for current status.

Leshoboro Plateau

Woodlot sites can provide interesting birdwatching environments. The Lesotho Woodlot Project began in 1973 with the aim of planting fast-growing trees like eucalyptus and fir for firewood and timber in many lowland areas. These woodlots are shown in the Lesotho map series 1: 50 000, and are thus fairly easy to locate. If you wish to visit a large number of these woodlots you should first contact the head office of the Project and obtain a permit. Great care should be taken not to damage fences and to close gates on the woodlots. I visited the woodlot site on the top of Leshoboro Plateau several times and observed many species including Grassbird, Wailing Cisticola, Redeyed Bulbul, Spotted Prinia and Cape White-eye. To reach the plateau, continue on the North Road past Marabeng Dam. At the second dam on the left-hand side after Marabeng you should park and walk up the hillside, through the wood in a south-westerly direction. When you see the first dam to the east, begin to look for a crevice in the wall – this will take you to the top of the plateau and into the main part of the woodlot. (Alternatively, there is a track for four-wheel-drive vehicles from the village of Letsetseng). I did not have time to make a survey of the whole area, but concentrated instead on walking westwards through an area of newly planted trees till I came upon an area of indigenous bush filled with many species of birds. I recommend such bush as well worth the effort of reaching it. Take care to return by the same route as there are only a few places where it is possible to get down from the top.

Nye-Nye Dam and Old Maputsoe Road Dam

Shortly before the junction to Maputsoe on the Maseru-Leribe road, is a dirt road to the left going to Ha Nye-Nye (and to the right a road to Mapoteng). Nye-Nye Dam is found at this junction. It is a fine birding spot. I have seen Little, Yellowbilled and Great White Egrets, African Spoonbill, South African Shelduck, Cape Shoveller, Whitefaced Duck, Purple Gallinule and Blackwinged Stilt. You can park on the dirt road and walk along the dam.

Not far from Nye-Nye Dam is another dam worth visiting, situated on the old Leribe-Maputsoe road (figure 3). To reach it, turn left off the Maseru-Leribe road onto the tarred road to Maputsoe. About 2 km along is the Khomokhoane rural development project. Turn sharp right onto a dirt road. The dam (Old Maputsoe Road Dam in the Checklist) is 1,5 km along on the left. This road can be impassable in the rains, so take care.

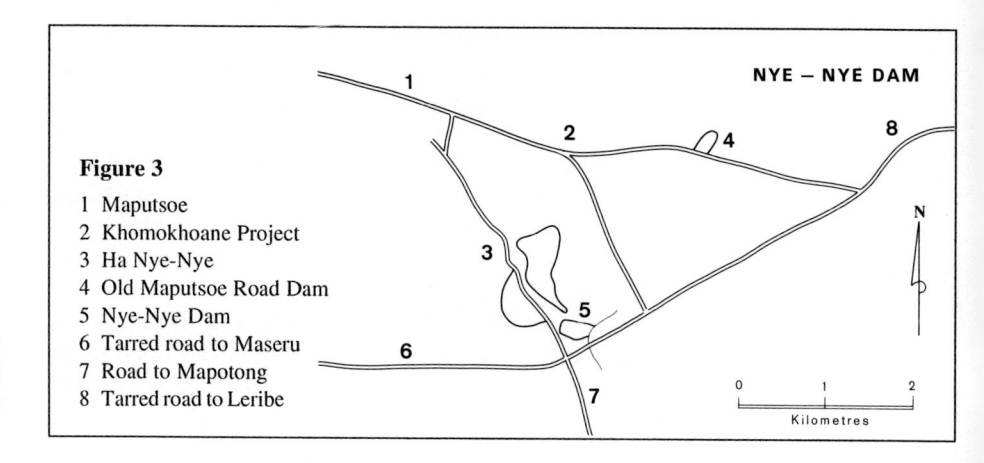

Figure 3
1 Maputsoe
2 Khomokhoane Project
3 Ha Nye-Nye
4 Old Maputsoe Road Dam
5 Nye-Nye Dam
6 Tarred road to Maseru
7 Road to Mapotong
8 Tarred road to Leribe

Little Egret, Bald Ibis, African Spoonbill and Eastern Redfooted Kestrel have been seen here. If you follow this road you will meet the main road again near St. Monica's Mission.

Leribe Dam

I have no name for this last dam, so I have called it Leribe Dam in the Checklist. It is just north of Leribe between the old and new roads joining Leribe and Butha-Buthe. It is a haven for wildlife. Along the wall of the dam is a narrow reed-bed where I have seen Malachite Kingfisher. Depending on the water level, many ducks can be seen on the water, as well as Ethiopian Snipe and other waders.

A trip to the dams mentioned will usually give you a sight of almost all herons, many ducks and waders, and with luck a few surprises as well.

South from Maseru

Take the Masianokeng road from the Circle out of Maseru (the road on the right of the Cathedral).

Morija Dam

Where the road from Morija meets the Maseru-Mafeteng road, there is a dam which is a good birding spot. There is a fence around it which you should take care not to damage. Along the wall of the dam, on the west side, and along part of the south side is a narrow fringe of reeds and at the east end is a grove of large

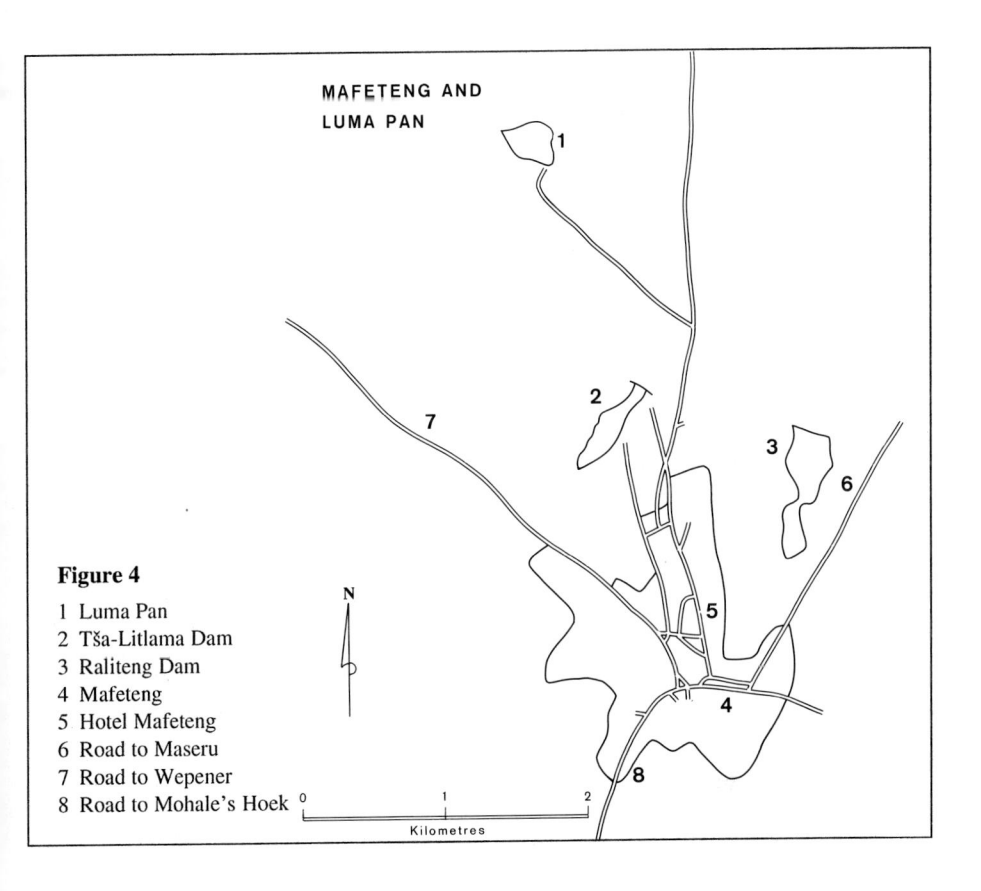

MAFETENG AND LUMA PAN

Figure 4
1 Luma Pan
2 Tša-Litlama Dam
3 Raliteng Dam
4 Mafeteng
5 Hotel Mafeteng
6 Road to Maseru
7 Road to Wepener
8 Road to Mohale's Hoek

trees. I have seen Darter, Blackcrowned Night Heron, African Spoonbill, Maccoa Duck, Marsh Sandpiper and Cape Reed Warbler here.

Mafeteng

Just north of Mafeteng (see figure 4) are two dams, Raliteng Dam, which you pass on your right just before arriving at Mafeteng, and Tša-Litlama Dam, which can be reached by following the road past the Hotel Mafeteng from the centre of town. The dam is on the left, beyond the hospital. A number of birds have been seen on Tša-Litlama: Whitebacked Duck, Kittlitz's Sandplover, other waders, Willow Warbler, and African Marsh Warbler. Raliteng Dam is not as good.

Luma Pan Carry on northwards on the road for Tša-Litlama past the Agricultural Department buildings and turn left where their gardens end. Luma Pan (see figure 4) is about 4 km from Mafeteng on this rather poor dirt road. In the older literature several species of birds have been reported from here. In the vicinity of Luma Pan, I have seen both Blue Korhaan and Pinkbilled Lark, neither of which is common, so even if Luma Pan itself is empty the area is generally well worth a visit.

Tša-Kholo To reach Tša-Kholo (see figure 5), which is the largest dam in the lowlands of Lesotho, take the tarred road from Mafeteng towards Wepener. After the hill, Qalabane, 14 km from Mafeteng, take a dirt road to the right, cross a stream and ascend a low ridge to the village of Ha Patsa. Approximately 23 km from Mafeteng, turn right, passing a store on your left side and cross the dam wall. At the eastern end of the wall the water flows over the road. If you continue instead of turning right, you come to another, smaller dam. From 1978 to 1980 Tša-Kholo received very little water so it was possible to walk the whole way around

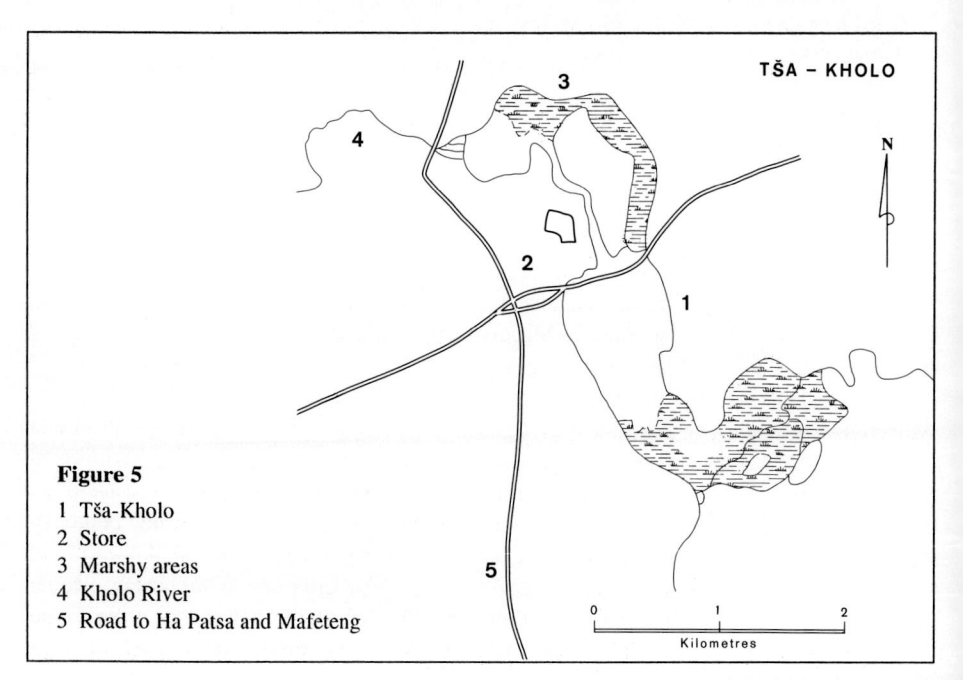

Figure 5
1 Tša-Kholo
2 Store
3 Marshy areas
4 Kholo River
5 Road to Ha Patsa and Mafeteng

the dam following the reeds and vegetation almost without getting one's shoes wet. In February 1981, after heavy rains, this was no longer possible. I have found the west side the best to walk along and usually spent a long time in this locality. Goliath Heron, Yellowbilled Egret and Avocet can be seen here. A visit from Maseru to Tša-Kholo I would regard as a whole day trip, starting early and arriving before it heats up too much. Returning to Maseru you can stop at Tša-Litlama and Morija Dams.

East from Maseru

National University of Lesotho (NUL) Roma

To reach NUL from Maseru take the Mafeteng road and turn left at Masianokeng, follow the road almost to its end (about 33 km from Maseru) and you will see the entrance on your left. The university is fenced and within this fence, apart from the teaching and residential areas, are some fields and a number of dams. The whole area is quite variable and as you can see from the Checklist, many different species have been seen on the campus over the years, so it is always worth paying a visit to NUL.

Koro-Koro

To reach Koro-Koro (figure 6) take the Roma road and approximately 22 km from Maseru, just after going sharply downhill, turn right onto a dirt road, which immediately crosses Mahlabathing Stream. After 28 km you come to Mokema Ha Rampoetsi, where you can see the beginning of the marshes on your right. Continuing, you pass St. Joseph's Roman Catholic Mission and about 1 km later the road crosses the reed beds of the marshes along a causeway, and turns sharply right through the village of Ha Phopheli. The road continues west till it meets the Maseru-Mafeteng road at the village of Ha Mantsebo. The Koro-Koro Marshes are the most extensive in Lesotho and many bundles of reeds are harvested there every year. From Mokema in the north they stretch about 5 km, past the Mission and up the Koro-Koro Valley. The northern parts are mostly meadow, interspersed with patches of reeds, where the cattle of the surrounding villages feed. Near St. Joseph's it becomes a dense reed-bed which continues up the valley. At Mokema Store a

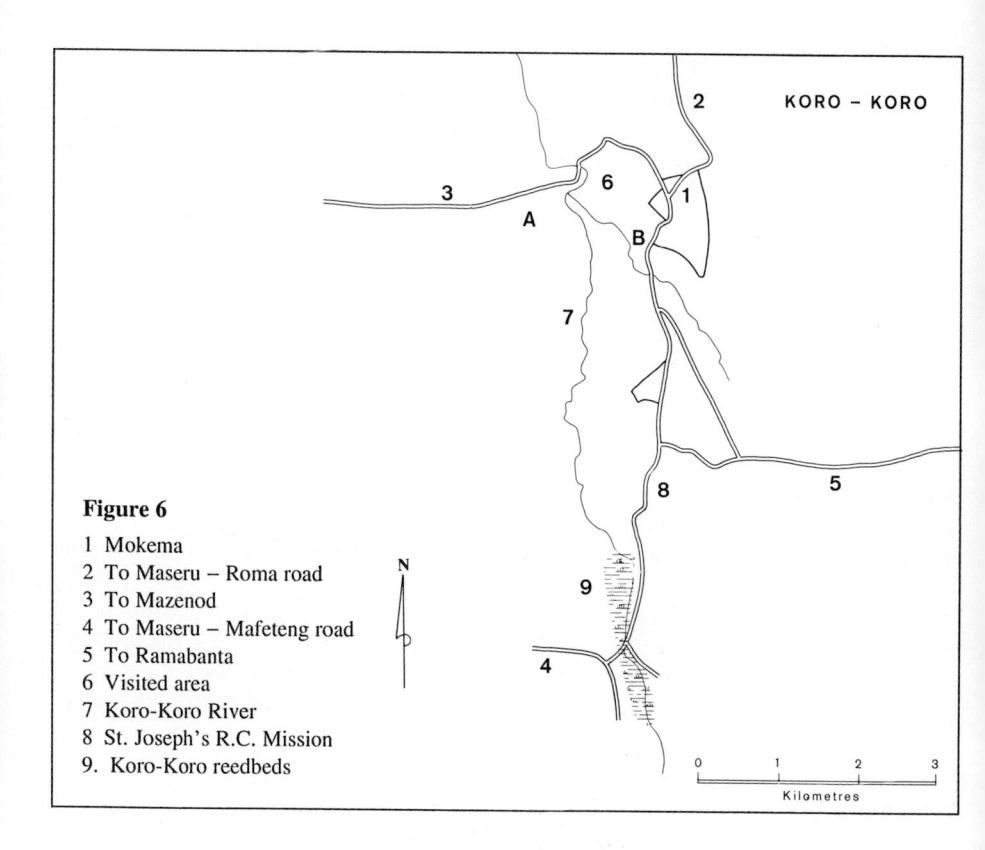

Figure 6
1 Mokema
2 To Maseru – Roma road
3 To Mazenod
4 To Maseru – Mafeteng road
5 To Ramabanta
6 Visited area
7 Koro-Koro River
8 St. Joseph's R.C. Mission
9. Koro-Koro reedbeds

rough road turns right off the road going south, encircling the northern end of the marshes. This is a feeder road for the Thaba Bosiu rural development project. If you follow it you come to the Maseru-Mafeteng road a little south of Mazenod. All my visits were to this northern end, and references to Koro-Koro in the Checklist are to this part. Usually being alone, I was not keen on entering the reed-beds at the southern end, but it is probably worth doing so, as birds like Little Bittern, African Rail, and African Crake may well be found there. From 1978 to 1980 the northern end had dried up to such an extent from the drought that it was possible to cross from A to B (see figure 6) without getting one's shoes wet. Through the whole period a variety of different species was recorded here: herons, ducks, waders and passerine birds. I would usually park after crossing the river that flows away from the marshes, and walk from A to B as far as possible. I would also follow the river downstream from the bridge, as the biotope here is quite different and recordings of Malachite Kingfisher, Rufousnaped Lark and Spotted Prinia are possible.

Outside the lowlands

Information on birdlife outside the lowlands is badly needed. Good birding areas are not as easily defined as in the lowlands (dams and woodlots). A good source of information on travelling in the mountains is David Ambrose's *Guide to Lesotho* as well as the 1: 25 000 and 1: 50 000 map series. You should always try to get the latest information on the state of the roads before travelling.

On the northern road

When I travelled from Maseru to Mokhotlong on the northern route I usually stayed overnight at New Oxbow Lodge (69 km from Butha-Buthe). The lodge is comfortable and can be used as a base for a longer stay (check with the Tourist Office for reservations). The road is usually good as far as Letšeng-la-Terae but it is wise to check locally, if you are not travelling in four-wheel-drive.

On the mountain road

Just north of Bushmen's Pass is a prohibited area as the Police Mounted Unit has a camp there. It is marked on official maps and is signposted so it is wise to avoid it, especially with a camera. After Bushmen's Pass you cross the Makhaleng River (56 km from Maseru). The area in the vicinity of the bridge is said to be a good birding place and it can be reached on a daytrip from Maseru. After a further 2 km, you come to Molimo Nthuse Hotel which can be used as a base for a weekend of walking and birdwatching. A visit from here to Qiloana Waterfall on the Makhaleng River can be recommended. Further along the road is Marakabei's (109 km from Maseru) where there is a Fraser Lodge, which makes a good base for a weekend trip to the mountains. Between Mantšonyane and Thaba-Tseka the country is very thinly populated and is really worth visiting. Nobody has done any recording here so far. The Thaba-Tseka area, especially, should be studied before the development of this, the newest regional headquarters, is too far advanced.

On the southern road

As one crosses the Senqu at Seaka Bridge, one passes from Mohale's Hoek district into Quthing district, and one can observe a marked difference in landscape and vegetation. Quthing has many trees, a contrast indeed to the stark dryness of Mohale's Hoek. The difference is said to be the result of past District Administrations' policy towards vegetation conservation.

In Quthing there is a hotel, which is a perfect base for birding in this part of the country. Going north out of Quthing the road takes you along the Senqu, giving you one of the most dramatic views in Lesotho, and in the red hot poker season, one of the most beautiful too.

At Mount Moorosi, the road leaves the Senqu to follow one of its major tributaries, the Quthing. The valley of this river comprises a rich variety of habitats and environments and has much to offer both the tourist and the ornithologist. The road eventually leads to Qacha's Nek via the large village of Mphaki, but the most interesting birdwatching venues lie in the vicinity of the Quthing. From the base of Ha Makoae, for example, one can explore valleys characterised by thick scrub and towering basalt cliffs. Fairy Flycatchers abound in the former, while Cape Vultures, Black Storks and Bearded Vultures soar over the latter. The headwaters of the Quthing rise in dramatic scenery dominated by basalt pinnacles, waterfalls and deep gorges, a truly remote region, little influenced by the outside world. The large dam of Letšeng-la-Letsie, near the Qoatlhamba (Drakensberg) watershed and about 96 km from Quthing, is easily accessible from the road to Ongeluksnek and is populated by a rich variety of birdlife: waders, wildfowl and other wetland species.

Gazetteer

Place names are entered in this list in the form in which they appear in modern sources.*
Variant spellings and previous names are cross-referenced.

Place Name	District	Map Reference
		S　　　　E
Agricultural Research Station	Maseru	29°17　27°30
Amebing Vlei *possibly* Raleting Dam		
Alwynskop	Quthing	30°25　27°38
Black Mountains	Mokhotlong	*c.* 29°31　29°12
Blue Mountain Pass	Maseru	29°26　27°58
Bokoaneng River (Bokong River (2), Little Bokong River)	Berea/Maseru	29°14　28°10 *to* 29°25　28°07
Bokong River (1)	Leribe/Thaba-Tseka	29°05　28°24 *to* 29°20　28°29
Bokong River (2) *see* Bokoaneng River		
Bushmen's Pass	Maseru	29°26　27°51
Butha-Buthe	Butha-Buthe	28°46　28°15
Caledon River (Mohokare River, NW boundary with OFS)	Butha-Buthe/Leribe/ Berea/Maseru/Mafeteng	
Corn Exchange	Leribe	28°58　27°59
Duma Pan *see* Luma Pan		
Fort Hartley	Quthing	30°20　27°45

23

Place Name	District	Map Reference
		S E
Ha Khitione	Maseru	29°30 27°44
Ha Maja	Maseru	29°25 27°38
Ha Makhate	Mohale's Hoek	30°04 27°25
Ha Makoae	Quthing	30°17 28°04
Ha 'Mamathe (Mamathe's)	Berea	29°07 27°50
Ha Marakabei (Marakabei's)	Maseru	29°33 28°09
Ha Ntsi	Maseru	29°24 27°46
Ha Patsa	Mafeteng	29°44 27°09
Ha Rampoetsi *see* Mokema Ha Rampoetsi		
Ha Sempe	Quthing	30°18 27°49
Ha Toloane	Maseru	29°38 27°28
Hellspoort (Hillsport)	Mafeteng	c. 29°52 27°24
Hensley's Dam *see* Nye-Nye Dam		
Hermon	Mafeteng	29°44 27°07
Hillsport *see* Hellspoort		
Hlotse *see* Leribe		
God-Help-Me Pass *see* Molimo-Nthuse Pass		
Jareteng River	Mokhotlong	29°20 29°27 *to* 29°17 29°23
Jordane River *see* Jorotane River		
Jorotane River (Jordane River)	Berea/Maseru	29°17 28°00 *to* 29°26 28°06
Kao River	Butha-Buthe	29°01 28°42 *to* 29°01 28°33
Khamolane Store	Berea	29°20 27°43
Khubelu River	Mokhotlong	28°46 28°52 *to* 29°17 28°53
Kolonyama	Leribe	29°04 27°44
Koma-Koma Bridge	Thaba-Tseka	29°36 28°43
Koro-Koro Marshes	Maseru	29°28 27°38
Koro-Koro River	Maseru	29°35 27°39 *to* 29°24 27°34
LAC = Lesotho Agricultural College *see* p.14		
Langalibalele Pass	Mokhotlong	29°17 29°25

Place Name	District	Map Reference	
		S	E
Langalibalele River (Lekhalabaletsi River)	Mokhotlong	29°16 29°16	29°26 to 29°21
Lehaha-la-Molapo	Leribe	29°07	28°20
Lekhalabaletsi River see Langalibalele River			
Leribe (Hlotse)	Leribe	28°52	28°03
Leribe Mission	Leribe	28°50	28°06
Leribe Dam	Leribe	28°51	28°03
Leshoboro Woodlot	Berea	29°14	27°36
Lesotho Agricultural College see p.14			
Letšeng-la-Letsie	Quthing	30°19	28°10
Letšeng-la-Terae	Mokhotlong	29°00	28°51
Likalaneng	Maseru	29°28	28°03
Linakeng River	Thaba-Tseka	29°36 29°33	29°06 to 28°45
Little Bokong River see Bokoaneng River			
Luma Pan (Duma Pan)	Mafeteng	29°47	27°14
Machache	Maseru	29°21	27°54
Mafeteng	Mafeteng	29°49	27°14
Mahlabatsaneng Pass	Mokhotlong	29°08	29°21
Makhalanyane	Maseru	29°25	27°37
Makhaleng Bridge	Mohale's Hoek	30°10	29°24
Makhaleng River	Maseru/Mafeteng/ Mohale's Hoek	29°20 30°19	27°57 to 27°23
Malaoaneng River = upper part of Bokoaneng River			
'Malefiloane Clinic	Mokhotlong	29°21	29°11
'Maletsunyane Falls	Maseru	29°52	28°03
Malibamatso River	Butha-Buthe/Leribe/ Thaba-Tseka	28°45 29°33	28°40 to 28°42
Mamathe's see Ha 'Mamathe			
Mantšonyane River	Thaba-Tseka	29°23 29°47	28°21 to 28°16
Mapholaneng	Mokhotlong	29°12	28°52
Maphotong Valley/Gorge	Maseru	29°28 30°02	27°45 27°40 to
Maphutseng River	Mohale's Hoek	30°19	27°27
Maputsoe	Leribe	28°54	27°54
Marabeng Dam	Maseru	29°16	27°36

Place Name	District	Map Reference
		S E
Marakabei's *see* Ha Marakabei		
Maseru	Maseru	29°19 27°29
Maseru West (Residential area in Maseru; *see* p.12)		
Masianokeng Woodlot	Maseru	29°23 27°33
Masite	Maseru	29°34 27°27
Masitise Mission	Quthing	30°24 27°39
Matsoku River	Leribe/Mokhotlong/ Thaba-Tseka	28°54 28°48 *to* 29°22 28°37
Mazenod	Maseru	29°25 27°33
Menoaneng Pass	Mokhotlong	29°25 28°57
Mkhotlong *see* Mokhotlong		
Mohale's Hoek	Mohale's Hoek	30°09 27°29
Mohlaka-oa-tuka	Maseru	29°26 27°46
Mohokare River *see* Caledon River		
Moiteling River	Mokhotlong	29°20 29°24 *to* 29°17 29°21
Mokema Ha Rampoetsi (Ha Rampoetsi)	Maseru	29°28 27°39
Mokhotlong *formerly* Mkhotlong	Mokhotlong	29°17 28°04
Mokhotlong River	Mokhotlong	29°24 29°26 *to* 29°16 29°02
Molimo-Nthuse Pass (God-Help-Me Pass)	Maseru	29°25 27°56
Mont-aux-Sources	Mokhotlong	28°46 28°52
Morija Dam	Maseru	29°37 27°30
'Moteng Pass	Butha-Buthe	28°45 28°36
Motete River	Butha-Buthe	28°54 28°48 *to* 28°59 28°33
Mountain Road	Maseru/Thaba-Tseka	29°25 27°41 *to* 29°32 28°37
Moyeni *see* Quthing		
Mphaki	Quthing	30°12 28°08
Mpharane	Mohale's Hoek	30°00 27°35
National University of Lesotho (NUL)	Maseru	29°27 27°43
New Europe (Residential area in Maseru; *see* p.12)		

Place Name	District	Map Reference	
		S	E
New Oxbow Lodge	Butha-Buthe	28°46	28°39
Nthlo-Kholo	Maseru	29°23	27°40
NUL see National University of Lesotho			
Nyakosoba	Maseru	29°31	27°46
Nye-Nye Dam (Hensley's Dam)	Leribe	28°54	27°56
Old Maputsoe Road Dam	Leribe	28°53	27°56
Ongeluk's Nek	Quthing	30°20	28°15
Orange River see Senqu River			
Outward Bound School	Leribe	29°00	28°09
Peka	Leribe	28°58	27°46
Pelaneng	Leribe	29°05	28°31
Phatlalla	Mohale's Hoek	30°23	27°34
Pitseng	Leribe	29°01	28°13
Polo ground (Maseru, see p.11)			
Qaba Lodge	Mafeteng	29°52	27°34
Qacha's Nek	Qacha's Nek	30°07	28°41
Qeme Plateau	Maseru	29°27	27°27
Qhoali River	Quthing	30°10 to 30°05	28°17 to 28°04
Qhomane Plateau	Maseru	29°32	27°37
Quthing (Moyeni)	Quthing	30°25	27°42
Raleting Dam	Mafeteng	29°48	27°15
Ramabanta	Maseru	29°40	27°48
Rampai's Nek	Leribe	28°47	28°13
Roma (St Peter's Village)	Maseru	29°28	27°44
Sani Pass	Thaba-Tseka	29°35	29°17
Sanqubethu = Sanqubetu	Mokhotlong	29°15 to 29°19	29°25 to 29°12
Seaka Bridge	Mohale's Hoek	30°22	27°34
Sebalabala	Berea	29°06	27°50
Sehlabathebe National Park	Qacha's Nek	29°52	29°08
Sekoati	Mafeteng	29°47	27°13
Semonkong	Maseru	29°50	28°03
Senqu River (Orange River)	Mokhotlong/Thaba-Tseka/Qacha's Nek/Quthing/Mohale's Hoek	28°53 to 30°19	29°01 to 27°23

Place Name	District	Map Reference
		S E
Senqunyane River	Berea/Thaba-Tseka/	29°08 28°16 *to*
	Maseru/Mohale's Hoek	30°02 28°10
Sewage ponds (Maseru, *see* p.13)		
Soosa	Maseru	29°22 28°00
St Michael's	Maseru	29°26 27°41
St Peter's Village (Roma)		
Tebetebeng Mill	Berea	29°07 27°47
Teyateyaneng	Berea	29°09 27°45
Thaba-Bosiu	Maseru	29°21 27°40
Thaba-Khupa Farm Institute	Maseru	29°23 27°37
Thaba-Phatšoa	Leribe	29°00 28°08
Thaba-Putsoa	Maseru	29°43 27°55
Thaba-Tseka	Thaba-Tseka	29°32 28°37
Thabana-li-'Mele	Maseru	29°36 27°47
Thabana-'Ntlenyana	Mokhotlong	29°28 29°16
Tša-Kholo	Mafeteng	29°40 27°10
Tša-Litlama Dam	Mafeteng	29°48 27°14
Tsatsa-le-Meno	Maseru	29°25 28°06
Tšehlanyane Dam	Butha-Buthe	28°44 28°37
Tsikoane Plateau and Woodlot	Leribe	28°55 28°01
Tsoelike River	Qacha's Nek	29°56 29°02 *to*
		29°59 28°41

UBLS = University of Botswana, Lesotho and Swaziland = National University of Lesotho

White City (Residential area in Maseru; *see* p.12)

Sources:

1. Map of Lesotho 1: 250 000, Series D.O.S. 621/1 Edition 1–D.O.S. 1978.
2. Map of Lesotho 1: 50 000, Series L50 (D.O.S. 421) Edition 4–D.O.S. 1979/1980.

Published for the Lesotho Government by the British Ministry of Overseas Development.

References
Birds of Lesotho

Contrary to the usual practice, I have arranged my references in two lists, one of material on the birds of Lesotho (as comprehensive as possible up to 1989), and one of general works consulted during the preparation of this book. Much of the information on the birds of Lesotho comes from mimeographed reports and unpublished manuscripts, and these I have annotated for the benefit of the reader.

Anon.	1965. Photographing the African Lammergeyer (*Gypaetus barbatus merionalis*) at the nesting site. *Oologists' Record* 39: 9–11.
Anon.	The record of a nine day horseback ride from Maseru to Sehlabathebe National Park, April 1977, made by a British volunteer.
	This manuscript was received from Victor Burke, 7 February 1984. It records the large and conspicuous birds seen daily during the ride, related to altitude, and was written at the request of Victor Burke. Unfortunately the name of the volunteer has been lost.
Aspinwall, A.M.	1973. Birds of the Campus and Roma Valley. Roma. 28 pp., mimeographed.
	This is the second list from Roma Valley. Together with the occurrence of each species, it gives a short description of the bird and also the Sesotho name for several species.
Balcomb, J.	Birds sighted in SNP as of June 1980. Compiled from the monthly diary of the park. Maseru. 7 pp.
	Unpublished list indicating month by month the occurrence of each species seen in the park from January 1975 to June 1980.

Boddam-Whetham, D.	1970. Lesotho birds in winter – a road count. *Bokmakierie* 22: 4–5.
Bonde, K.	1981. An annotated checklist to the birds of Lesotho. Preliminary edition. Maseru (Lesotho); Herning (Denmark). 87 pp., mimeographed.
	All relevant sightings by the author are published here.
Brickell, N.	1980. Lesotho bird counts during the years 1967–1977. Natal Avicultural Society. 9 pp., mimeographed. (Miscellaneous data on the keeping of cage and aviary birds: 1(1))
	Only species of interest to the aviculturist are mentioned.
Broekhuysen, G.J.	1955. White Stork (*Ciconia ciconia*) ringed in Poland and killed in Basutoland. *Ostrich* 26: 143.
Brown, C.J.	1988. A study of the Bearded Vulture *Gypaetus barbatus* in southern Africa. Ph.D. Thesis, University of Natal, Pietermaritzburg.
Brown, C.J. and Barnes, P.R.	1984. Birds of the Natal alpine belt. *Lammergeyer* 33: 1–13.
Brown, C.J., Brown, S.E. and Guy, J.J.	1988. Some physical parameters of Bearded Vulture *Gypaetus barbatus* nest sites in southern Africa. *Proceedings of the 6th Pan-African Ornithological Congress* 139–52.
Brown, C. and Whitfield, W.	1982. A Gymnogene kill. *Bokmakierie* 34: 81–2.
Brown, L.	Notes from Leslie H. Brown.
	Received from Victor Burke, 7 February 1984. Leslie Brown visited Lesotho with a team of consultants to write a development plan for tourism during the period September 1973 to February 1974. He broke his thigh falling from a horse, and stayed with Victor Burke while waiting for the plaster to be removed. His sightings are from early November 1973.
Burke, V.	The diary of Victor Burke, arranged in species order. Received 7 February 1984.
	Victor Burke stayed in Lesotho from December 1972 to 1978 at the Lesotho Agricultural College. As he was planning to do some work on the birds of Lesotho, he collected material, both published articles and unpublished manuscripts from birdwatchers, who were in Lesotho at the same time: H.G.M. Bass, David Halsted, Leslie H. Brown and N.A.L. Lexander. Victor Burke has kindly forwarded notes from those with whom I have not been in contact.

Carver, M.	1978. Lesotho. *Army Bird Watching Society Bulletin* 2: 1–2. Apart from a note on the sighting of Redshouldered Widow at Tsa-Khola, the article is an unspecified list of the birds seen on a 14-day trip to Lesotho by the author in March/April 1978.
Clancey, P.A.	1957. Notes on some birds from the highlands of Basutoland. *Ostrich* 28: 135–46.
Donnay, T.J.	1989. Semongkong Cape Vulture colony takes a plunge. *Bokmakierie* 41: 19–22.
Dove, R.	1971. Redbilled Firefinch (*Lagonosticta senegala*), new to Lesotho. *Ostrich* 42: 140.
Goodfellow, C.F.	Transcript from the diary of C.F. Goodfellow, lecturer at the University of Botswana, Lesotho and Swaziland, Roma, filed after his death in the library of the Percy FitzPatrick Institute of African Ornithology.
Goodfellow, C.F.	1966. Some additions to Jacot-Guillarmod's 'Catalogue of the birds of Basutoland (1963)'. *Ostrich* 37: 62–3.
Haagner, A.R.	1912. Migration report 1909–1911. *Journal of South African Ornithologists' Union* 7: 89–92.
Haagner, A.R.	1914. The White Stork in South Africa. *Journal of South African Ornithologists' Union* 9(1): 52–5. Reprint of the author's article in *Aquila* (1912).
Halsted, D.	1974. Interesting bird observations noted. *Linonyana tsa Lesotho* (Birds of Lesotho) 1(1).
Hean, A.R. and Mokhehle, N.C.	1946–47. Some Basuto beliefs about wild life. *African Wild Life* 6: 68–9 and 80–2.
Jacot-Guillarmod, C.	1932. Some notes on birds in Basutoland. *Ostrich* 3: 35–40.
Jacot-Guillarmod, C.	1955. Zoologists in Basutoland. *Basutoland Scientific Association Annual Report* 7–9.
Jacot-Guillarmod, C.	1963. Catalogue of the birds of Basutoland. The South African Avifauna Series, 8. Cape Town: Percy FitzPatrick Institute of African Ornithology. 111 pp., mimeographed. The first comprehensive list for Lesotho.
James, H.W.	1970. Catalogue of the birds' eggs in the collection of the National Museum of Rhodesia. Salisbury: Trustees of the National Museum of Rhodesia for the Queen Victoria Museum. 237 pp. One of the collections in the museum comes from J.A. Cottrell and was purchased in 1963. Most of the eggs are from Barotseland and Zambia, but there are also eggs from Lesotho. Most clutches were

	collected in the 1920s. Cottrell is mentioned as a collector by C. Jacot-Guillarmod in the *Basutoland Scientific Association Annual Report* (1956). As far as I can see, Cottrell is the only one who has collected eggs in Lesotho on any scale.
Jilbert, J.	197? Birdwatching in Lesotho: a series of booklets on areas of ornithological interest. Illustrated by Pauline Jilbert. Vol. 1: The Mountain Road. Roma. 19 pp., mimeographed. This is the first attempt to cover a certain area, giving instructions on how to reach it and mentioning some of the birds to be seen. Only Vol. 1 was issued.
Jilbert, J.	1979. Cape Vulture sites in Lesotho – a summary of current knowledge. *Vulture News* No. 2: 3–14.
Jilbert, J.	1982. 1982 Lesotho Cape Vulture project – preliminary report. *Vulture News* No. 8: 19–25.
Jilbert, J.	1983. Cape Vulture sites in Lesotho: a continuing investigation. *Bokmakierie* 35: 85–88.
Jones, P.J.	1978. A possible function of the 'wingdrying' posture in the Reed Cormorant (*Phalacrocorax africanus*). *Ibis* 120: 540–42. The study took place at the Agricultural Research Station, Maseru, in 1977.
Lexander, N.A.L.	1977. Lesotho Birds, Maseru. N.A.L. Lexander stayed in Lesotho from 1973 to about 1980, working for FAO, and the notes received from him cover only the first half of his stay. Victor Burke never discussed them with Lexander, and I have not been able to reach him, so I have for once dropped my own principle of recording everything, as there are a few sightings of Lexander's which I would like to discuss with him before they are published.
Maclean, G.	1955. House Sparrows spreading into Basutoland. *Ostrich* 26: 46.
MacLeay, K.N.C.	1970. An annotated list of the birds observed in the Roma area of Lesotho: August 1964 to May 1969. 15 pp., mimeographed. This is the first list covering the Roma area.
Manry, D.E.	1985. Distribution, abundance and conservation of the Bald Ibis *Geronticus calvus* in southern Africa. *Biological Conservation* 33: 351–62.
Mendelsohn, J.	1984. The Mountain Pipit in the Drakensberg. *Bokmakierie* 36: 40–44.
Murray, J.P.	Some MS notes of J.P. Murray. 9 pp. These notes are extracted from copies of Layard & Sharpe's *Birds of South Africa* (1875–84) and Stark & Sclater's *The birds of South*

	Africa, Vols. 1–4, (1900–1906) formerly in the possession of Mr. J.P. Murray, who was a civil servant in Basutoland (*c.* 1894–1926), staying in Mafeteng and Maseru. The notes are arranged in the order used by C. Jacot-Guillarmod in his *Catalogue*.
Murray, J.P.	1909a. Marked White Stork in Basutoland. *Journal of South African Ornithologists' Union* 5: 115–16.
Murray, J.P.	1909b. Allen's Gallinule in Basutoland. *Journal of South African Ornithologists' Union* 5: 116.
Murray, J.P.	1914. Spotted Crake in Basutoland. *Journal of South African Ornithologists' Union* 9: 65.
Murray, J.P.	1964. Some MS notes on Basutoland birds. Edited by J.M. Winterbottom. *The South African Avifauna Series, 21*. Cape Town: Percy FitzPatrick Institute of African Ornithology. 10 pp., mimeographed.
	This is a small collection of the notes J.P. Murray made in his 'Stark & Sclater' and 'Layard', only published after the *Catalogue* of Jacot-Guillarmod to fill gaps in the text.
Passineau. L.	1978. Birds of the Sehlabathebe National Park. Maseru: Parks Administration, Lesotho National Parks. 27 pp., mimeographed.
	A locality guide, introducing the general public to birdstudy, plus an identification key, a checklist, and a list of Sesotho bird names.
Quickelberge, C.D.	1972. Results of two ornithological expeditions to Lesotho. *Durban Museum Novitates* 9: 251–78.
Reichardt, M.	1980. *Tourist guide to birds of Lesotho, with bird names in seven languages*. Roma: ESSA.
Rudebeck, G.	1956–63. Observations of the Bearded Vulture (*Gypaetus barbatus*) in South Africa; Studies on some Palaearctic and Arctic birds in their winter quarters in South Africa, parts 1–5. In Hanström, B., Brinck, P. and Rudebeck, G. (eds) *South African Animal Life: Results of the Lund University Expedition in 1950–51*. 4: 406–415, 459–498; 9: 418–516. Stockholm: Almqvist & Wiksell.
Steele, P.H.	1976. Checklist of the more common birds species found at Thaba-Khupa Farm Institute.
	A list of the birds Philip Steele saw during his stay at the Farm Institute, unfortunately without any indication of the commonness of each species.
Symons, R.E.	1916. Nesting of the Black Stork (*Ciconia nigra*). *Journal of South African Ornithologists' Union* 11: 148–49.

Symons, R.E.	1919. Birdlife in the Drakensberg, Natal and Basutoland. *South African Journal of Natural History* 1: 224–38.
Symons, R.E.	1920. Birdlife in the Drakensberg. The Black Eagle. *South African Journal of Natural History* 2: 253–56.
Van der Plaat, A.	1952. House Sparrow (*Passer domesticus*). *Ostrich* 23: 64.
Vincent, J.	1947–48. New races of a Tit-babbler and a Lark from the Basutoland Mountains. *Bulletin of the British Ornithologists' Club* 68: 145–46.
Vincent, J.	1951. The description of a new race of Richard's Pipit *Anthus richardi* (Vieillot) from Basutoland. *Annals of the Natal Museum* 12: 135–36.
Vincent, J.	1950. New races of the Cape Bunting from Southern Rhodesia and Basutoland. *Bulletin of the British Ornithologists' Club* 70: 14–17

During the preparation of this book I have corresponded with several persons who have kindly sent me their own sightings and I wish to express my gratitude to the following:

David Ambrose, H.G.M. Bass, Christopher Brown, Clive C. Clements, Dave Coghlan, Digby Cyrus, Donald Davidson, Canon Reginald Dove, David Halsted, John Jilbert, P.J. Jones, L.G.A. Smits, Walter Stanford, Godfrey Symons, John Williamson, Jack Vincent.

General references

Ambrose, D.	1976. *The guide to Lesotho.* Johannesburg: Winchester Press.
Ambrose, D.	1984. Lesotho. In *The guide to Botswana, Lesotho and Swaziland.* Saxonwold: Winchester Press.
Ambrose, D.	1983. *Lesotho's heritage in jeopardy: Report of the Chairman of the Protection & Preservation Commission for the years 1980–81 and 1982–83, together with a survey of its past work and present challenges.* Maseru: The P.P.C.
Brooke, R. K.	1984. *South African red data book – Birds.* South African National Scientific Programmes Report, No. 97. Pretoria: Foundation for Research Development.
Bureau of Statistics. Kingdom of Lesotho	1977–82. *1976 Population census report.* Maseru: The Bureau of Statistics.
Burrow, J.	1971. *Travels in the wilds of Africa: being the diary of a young scientific assistant, who accompanied Sir Andrew Smith in the Expedition 1834–36.* Edited with notes and index by Percival R. Kirby. Cape Town: A. A. Balkema.
Clancey, P. A. (Ed.)	1980. *SAOS checklist of southern African birds.* Johannesburg: SAOS.
Cyrus, D. and Robson, N.	1980. *Bird atlas of Natal.* Pietermaritzburg: University of Natal Press.
Hall, B. P. and R. E. Moreau	1970. *An atlas of speciation of African passerine birds.* London: British Museum (Natural History).

Jacot Guillarmod, A.	1971. *Flora of Lesotho (Basutoland)*. Lehre, BRD: J. Cramer.
Layard, E.L.	1875–84. *Birds of South Africa*. 2nd ed. Revised by R. Bowdler Sharpe. London.
Leistner, O.A. and Morris, J.W.	1976. *Southern African place names*. Grahamstown: Cape Provincial Museum and Albany Museum. (Annals of the Cape Provincial Museum, Vol.12).
Lye, W.F.	1975. *Andrew Smith's journal of his expedition into the interior of South Africa, 1834–36 : an authentic narrative of travels and discoveries, the manners and costumes of the native tribes, and the physical nature of the country*. Introduction and notes by W.F. Lye. Contemporary illustrations by Charles Davidson Bell. Cape Town: A.A. Balkema for the South African Museum.
Mackworth-Praid, C. and Grant, C.H.B.	1962–63. *Birds of the southern third of Africa*. Vols.1–2. London: Longmans.
McLachlan G.R. and Liversidge, R.	1978. *Roberts birds of South Africa*. 4th ed. Cape Town: John Voelcker Bird Book Fund.
Maclean, G.L.	1985. *Roberts' birds of southern Africa*. 5th ed. Cape Town: John Voelcker Bird Book Fund.
Mundy, P.J.	1978. The Egyptian Vulture (*Neophron percnoterus*) in southern Africa. *Biological Conservation* 14: 307–315.
Sharpe, R.B.	1906. Birds. In *The history of the collections contained in the Natural History Departments of the British Museum*. Vol.2. London: British Museum (Natural History).
Skead, C.J.	1967. *The sunbirds of southern Africa, also the sugarbirds, the white-eyes and the Spotted Creeper*. Assisted by Cecily M. Niven, J.M. Winterbottom and Richard Liversidge. Cape Town: A.A. Balkema, published for the Trustees of the South African Bird Book Fund.
Smith, A.	1939–40. *The diary of Andrew Smith, director of 'Expedition for Exploring Central Africa', 1834–36*. Edited, with an introduction, footnotes, map, and indexes by Percival R. Kirby. Cape Town: The Van Riebeeck Society.
Smith, Andrew	1977. *Illustrations of the zoology of South Africa*, Vols.1–3. Fascimile reprint of the original work published in London during 1849 with an introduction by R.F. Kennedy. Johannesburg: Winchester Press.

Smith, Andrew	1836. Report of the expedition for exploring Central Africa under the superintendence of Dr A. Smith. *Journal of the Geographical Society of London* 6: 394–413.
Snow, D.W.	1978. *An atlas of speciation of African non-passerine birds.* London: British Museum (Natural History).
Stark, A.C. and Sclater, W.L.	1900–06. *The birds of South Africa*, Vols.1–4. London: R.H. Porter.
Willet, S.M.and Ambrose, D.	1980. *Lesotho: a comprehensive bibliography.* Oxford: Clio Press. (World Bibliography Series, 3)

Notes on the checklist

The order of the Checklist and its nomenclature are from the fifth edition of *Roberts' birds of southern Africa*, by Gordon Lindsay Maclean (Cape Town: John Voelcker Bird Book Fund, 1984).

As it is difficult to determine the subspecific status when most of the material consists of field notes and only a small fraction consists of skins, I have used only the binomial scientific form rather than the trinomial. Only subspecies that can easily be identified in the field, such as *Lanius collaris subcoronatus*, are mentioned.

Generally I have used five categories to describe the status of a certain species. These categories might have to be changed in the future, when more recent material has been collected.

Hypothetical — One or a few very old and often unspecified records, all older than 1930.

Single record — So far only one known record of this species since 1930. Older records disregarded.

Rare — A few records only, some of which might even be unspecified (e.g. without date, number of birds seen).

Uncommon — Encountered regularly, but infrequently.

Common — Encountered regularly and frequently.

Analysis of status categories

Status category	Checklist	Preliminary edition
Hypothetical	21	30
Single record	39	33
Rare	83	67
Uncommon	40	48
Common	119	110
Total	302	288

Annotated checklist

Family   PODICIPEDIDAE: Grebes

Two species recorded in Lesotho.

Great Crested Grebe
Podiceps cristatus
006

Apart from one record from Sehlabathebe National Park, January 1980 (Balcomb MS), the only records are from Mafeteng at the beginning of the century (Murray MS). Single record.

Dabchick
Tachybaptus ruficollis
008

Common resident breeding bird in the lowlands on most small to medium-sized dams. Not seen yet at Tša-Kholo. In winter often seen in numbers of 20 or more together.

Family   PHALACROCORACIDAE: Cormorants

Two species recorded in Lesotho.

Whitebreasted Cormorant
Phalacrocorax carbo
055

Symons (1919) found this species breeding commonly in Mokhotlong Valley, in September 1914. This is the only breeding record. There is no information on its present status in the mountains, apart from sightings at Kao, Malibamatso River in April 1972 (Lexander MS) and Tšehlanyane Dam, Malibamatso, early November 1973 (Brown MS). Recorded in the lowlands throughout the year in numbers fewer than five; not as common as *P. africanus*.

Reed Cormorant
Phalacrocorax africanus
058

Common in the lowlands, and can be seen throughout the year on all medium stretches of water. No breeding records. No records yet from Tša-Kholo. Might be found outside the lowlands, but more material is needed to determine its actual status there.

Family

ANHINGIDAE: Darters

One species in Lesotho.

Darter
Anhinga melanogaster
060

Uncommon in Lesotho; a few records from Roma (Aspinwall 1973). Maseru, Agricultural Research Station (Bonde 1981, Davidson, D. *in litt.*); Tša-Kholo (Halsted, D. *in litt.*) and Morija (Jones, P.J. *in litt.*). Occasional in the mountains (Jacot-Guillarmod 1963). Noted from Kao, April 1972 (Lexander MS).

Family

ARDEIDAE: Herons, egrets and bitterns

Twelve species in Lesotho.

Grey Heron
Ardea cinerea
062

Common in the lowlands. Can be seen in the mountains as high as in the mountain region. Normally seen singly, and always close to water, which limits its distribution. Breeding recorded: near Khamolane Store, Berea Plateau, (Burke MS); at Masitise Mission, Quthing (Brown MS); Sanqubetu Valley, Mokhotlong district (Symons, G. *in litt.*). More material is however needed to determine the size of the breeding population.

Blackheaded Heron
Ardea melanocephala
063

A common heron in Lesotho both in the lowlands and in the mountains. Can be seen in Mokhotlong district as high as in the mountain region. It seems independent of water. Breeding recorded near Khamolane Store, Berea Plateau, November 1973 – 20 to 30 nests with feathered young. This place is known to have been in use from at least 1967 (Burke MS). Breeding suspected near Teyateyaneng (Brown, C. *in litt.*). More material is however needed to determine the actual size of the breeding population.

Goliath Heron
Ardea goliath
064

Uncommon if not rare, found only in the lowlands. Roma, January 1965 (Macleay 1970); Tša-Kholo, January 1974 (Halsted, D. *in litt.*); Tša-Kholo, January 1979 and at Morija, September 1980 (Bonde 1981); Leribe Dam, January 1982 (Davidson, D. *in litt.*).

Purple Heron
Ardea purpurea
065

Common but less numerous than *A. cinerea* and *A. melanocephala* in the lowlands. It is most often seen when flushed from reeds and bushes over water, so may be overlooked. No breeding records, and no records from outside the lowland region.

Great White Egret
Egretta alba
066

Both Murray (MS) and Jacot-Guillarmod (1963) occasionally met with this rare species. In recent years eight records have been collected: Koro-Koro, January 1967 (MacLeay 1970); Agricultural Research Station, January 1974 (Burke MS); at Pelaneng, Malibamatso River, April 1972 (Lexander MS); Nye-Nye Dam, February 1979; Koro-Koro, March 1979, October 1979, November 1979 (Bonde 1981); Agricultural Research Station, November 1982 (Davidson, D. *in litt.*). The Pelaneng record (upper mountain valley region) might indicate that this species could be spreading into the mountains along the rivers.

Little Egret
Egretta garzetta
067

Apart from an undated skin in the Transvaal Museum, collected by A.J. Moffatt 'Caledon River', this species is first mentioned by Aspinwall (1973). Also reported by Burke (MS) and Lexander (MS). It is not noted by Bass 1970–73 (*in litt.*). From

1978 to 1980, Bonde (1981) has at least 17 records from various dams in the lowlands, from all months of the year. Has this species become more common in recent years? It seems to be the most common of the three bigger white egrets. Common in the lowlands.

Yellowbilled Egret
Egretta intermedia
068

The rarest of the white egrets in Lesotho. Apart from a record from the mountains near Mokhotlong, December 1947 (Vincent, J. *in litt.*), it has been reported ten times from the lowlands.

Cattle Egret
Bubulcus ibis
071

Has become more common throughout the century. Murray (MS) stated that it visited Basutoland now and again. Breeding reported at: Marabeng Dam, October 1973 – about 30 nesting in willows; and October 1974 – about 24 nestbuilding (Burke MS). Breeding also reported in gum forest near Teyateyaneng early November 1973 – about 100 pairs (Brown MS). I have not been able to locate this gum forest but a qualified guess is that it is at Tebetebeng Mill a little north of Teyateyaneng on the road to Ha Mamathe. Very common in the lowland region, but no records from other parts of the country.

Squacco Heron
Ardeola ralloides
072

Rare. Recorded seven times: at the University, Roma, 27 January 1965 (Goodfellow 1966); Leribe Dam, 6 December 1968 (Dove, R. *in litt.*); Maseru, January 1971 (Bass, H.G.M. *in litt.*); Agricultural Research Station, 18 February 1973; New Europe, Maseru, 13 October 1973 (Lexander MS); Agricultural Research Station, 10 April 1980, and 14 May 1983 (Davidson D. *in litt.*).

Blackcrowned Night Heron
Nycticorax nycticorax
076

In recent years seen commonly in the Maseru-Roma area, and at Morija. No breeding records so far, though birds in juvenile plumage often seen.

Little Bittern
Ixobrychus minutus
078

Very rare. Davidson (*in litt.*) has two records, both from Agricultural Research Station: 19 January 1981, 2 March 1983. Bonde (1981) also has three records from the Agricultural

Research Station, all at lake 2: 26 November 1978, flew from one reedbed to another; 23 December 1979, 18 January 1981, one seen sitting at the edge of the reedbed at the far end of the lake, next to the fishponds.

Bittern
Botaurus stellaris
080

Murray (MS), stated that he shot one near Maseru, December 1910, saw a few more in different parts of Basutoland, and heard them booming several times. *Khuiti oa Mahlaka* to the Basuto, literally 'mole of the reeds'. Jacot-Guillarmod (1963) mentioned it in parenthesis. With no recent records it must have declined sharply, and should not today be regarded as a bird of Lesotho.

Family

SCOPIDAE: Hamerkop

One species in Lesotho.

Hamerkop
Scopus umbretta
081

Common in the lowlands throughout the year; not so numerous outside the lowland region. Breeding recorded from NUL, Roma, (Aspinwall 1973); Likhaleng (Bass, H.G.M. *in litt.*); Mokhotlong, and Maphotong Gorge (Burke MS).

Family

CICONIIDAE: Storks

Four species in Lesotho.

White Stork
Ciconia ciconia
083

Common summer visitor, from September to late March, to be met with all over the country.

Black Stork
Ciconia nigra
084

Can be seen throughout the country. Breeding reported from Mokhotlong district (Vincent, J. *in litt.*); and from Roma (MacLeay 1970, Aspinwall 1973). Size of breeding population not known.

Abdim's Stork
Ciconia abdimii
085

Common summer visitor to the lowlands, where flocks of various sizes can be seen, often in association with *Ciconia ciconia*. So far no records have been received from other parts of the country.

Yellowbilled Stork
Mycteria ibis
090

Uncommon visitor to the lowlands. Jacot-Guillarmod (1963) states that 'this species is a common summer visitor in the lowlands, usually seen in small parties of about five, very often birds with immature plumage are present'. From more recent material it is apparent that it might have decreased in numbers.

Family

PLATALEIDAE: Ibises and spoonbills

Five species recorded in Lesotho.

Sacred Ibis
Threskiornis aethiopicus
091

Common in the lowlands, but not as common as *Bostrychia hagedash*. Can also be seen from November to March in the southern parts of the mountains: Sehlabathebe (Balcomb MS); Letšeng-la-Letsie (Brown MS, Lexander MS, Bonde 1981).

Bald Ibis
Geronticus calvus
092

Vagrant to the lowlands outside the breeding season. It seems still to be common. Recent material from nesting sites necessary to determine its actual status. It is possible that numbers may be decreasing.

Glossy Ibis
Plegadis falcinellus
093

Rare. Recorded four times in Lesotho: Koro-Koro, 18 July 1970 (Ambrose, D. *in litt*.); Koro-Koro, 18 March 1979; dam next to White City, Maseru, 9 September 1979 (Bonde 1981); Agricultural Research Station, 16 March 1982 (Davidson, D. *in litt*.).

Hadeda Ibis
Bostrychia hagedash
094

Has increased in recent years, and is today common in the lowland region throughout the year. Can also be seen in the upper mountain valley region in small numbers. May be spreading into the mountains following the rivers. So far no breeding records.

African Spoonbill
Platalea alba
095

Fairly common in the lowlands throughout the year, normally seen singly or a few together. Outside the lowland region there is only one record: Letšeng-la-Letsie, 6 March 1981 (Bonde 1981). No breeding records.

Family

PHOENICOPTERIDAE: Flamingoes

Two species in Lesotho.

Greater Flamingo
Phoenicopterus ruber
096

Single record. This species has been recorded once: Tša-Kholo, 10 March and 22 March 1985 – a flock of 38 (Ambrose, D. *in litt.*).

Lesser Flamingo
Phoenicopterus minor
097

Hypothetical. Not recorded in recent years; the only record is Luma Pan, Mafeteng, 1901 (Murray MS).

Family

ANATIDAE: Swans, geese and ducks

Sixteen species recorded in Lesotho, four species breeding.

Whitefaced Duck
Dendrocygna viduata
099

A newcomer to Lesotho with a rather odd distribution, as it has only been seen around Maseru. First recorded in 1967 by Brickell (1980) at the Agricultural Research Station. In 1974 breeding was recorded at the Agricultural Research Station – pair with ducklings (Brickell 1980). Breeding also recorded at the same locality in 1976 (Burke MS) and in 1977 (Jones, P.J. *in litt.*). Commonly seen at Agricultural Research Station 1978, 1979, and 1980. Also found breeding at sewage ponds behind railway station, Maseru, in 1981. One sighting from Leribe district, Nye-Nye Dam, 4 February 1979 (Bonde 1981).

Fulvous Duck
Dendrocygna bicolor
100

Rare. Tša-Kholo in 1967. In 1969, about 15 flying out from Tša-Kholo in the direction of Mafeteng. Seen two days later, also about 15, at Luma Pan (the same group?). Tša-Kholo in

1970 and 1974. Seaka Bridge area, 1975 – about a dozen (Brickell 1980). Maseru, Caledon River, 15 March 1980 (Davidson, D. *in litt.*). Apart from one Maseru record, only recorded in the southern part of the country.

Whitebacked Duck
Thalassornis leuconotus
101

A newcomer to Lesotho. Apart from an old record from Mafeteng, 1910 (Murray MS), the first record is from Leribe Dam, 22 June 1968 (Dove, R. *in litt.*). It is most common in the northern part of the lowlands, south of Maseru. Recorded during all months of the year, but uncommon. So far no breeding records.

Egyptian Goose
Alopochen aegyptiacus
102

Common in the Maseru-Roma area, normally singly or only a few together. Breeding recorded at the Agricultural Research Station, 26 November 1978 (Bonde 1981). One sighting north of Maseru District, Leribe Dam, 1 August 1978 (Bonde 1981). More sightings from the south of the country: Letšeng-la-Letsie, early November 1973 (Brown MS); 6 March 1981 (Bonde 1981). Reported from Sehlabathebe National Park throughout the year, except April and October (Balcomb MS). Old reports that it was common on the Mokhotlong River (R. Symons 1919), are not substantiated by more recent information.

South African Shelduck
Tadorna cana
103

An uncommon visitor to the south-west of the country, normally in small numbers, rare in the northern lowlands. No records from the interior mountains. There are two interesting sightings of large flocks from Letšeng-la-Letsie: early November 1973 (Brown MS) and 6 March 1981 (Bonde 1981).

Yellowbilled Duck
Anas undulata
104

The most common breeding duck in the lowlands on dams, seen throughout the year. Common in Sehlabathebe National Park, where it has been noted in all months of the year (Balcomb MS). Also Letšeng-la-Letsie, early November 1973 (Brown MS).

African Black Duck
Anas sparsa
105

Common on most rivers both in the lowlands and in the mountains, only occasionally seen on dams. Breeding presumed, but so far only one record: Sanqubetu Valley, 1 December 1968 – seen with large young (Symons, G. *in litt.*).

Cape Teal
Anas capensis
106

Very rare in Lesotho. So far three records only: Mafeteng, 1901 (Murray MS); Caledon and sewage ponds, Maseru, 24 September 1973 (Halsted 1974); Koro-Koro, September 1976 (Jilbert J. *in litt.*).

Hottentot Teal
Anas hottentota
107

Apart from three birds shot around the turn of the century (Murray MS), there is only one record: Agricultural Research Station, 22 to 26 June 1977 (Jones, P.J. *in litt.*).

Redbilled Teal
Anas erythrorhyncha
108

Common in the lowland region, but not as numerous as *A. undulata*. There is only one record from the mountains: Sehlabathebe National Park, January 1980, (Balcomb MS). Breeding reported: Agricultural Research Station, 12 March 1980; Nye-Nye Dam, 25 January 1981 (Bonde 1981).

Cape Shoveller
Anas smithii
112

The editor of Murray (1964), J.M. Winterbottom, has noted of this species: 'Taken in conjunction with Jacot-Guillarmod and what is known of the bird's movements elsewhere, it would appear that the Shoveller is an uncommon passage migrant in Basutoland'. The more recent material seems to provide confirmation for this picture: Koro-Koro, August 1966 (MacLeay 1970); Agricultural Research Station, 1 December 1973 (Halsted, D. *in litt.*); 18 January 1981 (Bonde 1981); 10 March 1983 (Davidson, D. *in litt.*); Nye-Nye Dam, 1 August 1978, 11 March 1979; dam on old Maputsoe road, 4 February 1979; Morija Dam, 30 September 1979; Tša-Kholo, 7 January 1979, 4 February 1981; Letšeng-la-Letsie, 6 March 1981 (Bonde 1981).

Southern Pochard
Netta erythrophthalma
113

Fairly common in the lowlands, except in late summer and fall, but apart from one record at the Agricultural Research Station, 13 September 1980, of about 35 male and 20 female (Bonde 1981) it is not numerous. Breeding not recorded.

Pygmy Goose
Nettapus auritus
114

There is only one old record of one shot near Mafeteng, 1901 (Murray MS). It can only be regarded as hypothetical in Lesotho.

Knobbilled Duck
Sarkidiornis melanotos
115

Very rare in Lesotho; only five records in the lowlands from 1930 up to 1980.

Spurwinged Goose
Plectropterus gambensis
116

Seems to have decreased in numbers since the beginning of the century, when it was common and also found breeding (Murray MS). Since then there have been no breeding records. It must be regarded as an uncommon summer visitor to the lowlands.

Maccoa Duck
Oxyura maccoa
117

Fairly common in the lowlands, where it can been seen on most dams in all months of the year. So far no breeding records.

Family

SAGITTARIIDAE: Secretarybird

One species in Lesotho.

Secretarybird
Sagittarius serpentarius
118

Widely distributed in the lowlands, but rare (Jacot-Guillarmod 1963). The more recent material supports this statement. It is also reported from Sehlabathebe National Park in all months (Balcomb MS). No breeding records.

Family

ACCIPITRIDAE: Eagles, hawks, buzzards, etc.

Twenty species recorded in Lesotho, about half only a few times.

Bearded Vulture
Gypaetus barbatus
119

Lesotho is the last stronghold of this species in southern Africa. It is resident and breeding, and still fairly common in the mountains. Brown (1988) has estimated the total adult population in southern Africa to be 203 pairs. Of these 122 breed in Lesotho.

Egyptian Vulture
Neophron percnopterus
120

One killed near Mamathe's after a hailstorm, around 1930 (Jacot-Guillarmod 1963). Has decreased sharply in southern Africa since the beginning of the century. Its present status in southern Africa is discussed by Mundy (1978). The post-1945 sightings of this species show that nine of 19 birds have been seen in Transkei and the question is raised, but not answered, whether the vulture still breeds in South Africa (Transkei). It might not be extinct, but must be regarded as hypothetical in Lesotho.

Cape Vulture
Gyps coprotheres
122

In Lesotho there are three known breeding colonies and five places with suspected breeding colonies (Jilbert 1979, 1983). The number of breeding pairs was estimated to be 90 by Brooke (1985). Since then it seems that there might have been some decline in the numbers as the colony at Semongkong has declined sharply (Donnay 1989). If the described decline is general for Lesotho, then the number of breeding birds must be much lower now. There are however still parts of the country which have not been investigated in depth, so it is difficult to give an accurate figure.

Black Kite Yellowbilled Kite
Milvus migrans
126

Formerly this species was separated into two different species: Yellowbilled Kite, *Milvus parasitus* and Black Kite, *Milvus migrans*. *Milvus m. migrans* is a non-breeding visitor to southern Africa from the Palaearctic, and has been recorded once in Lesotho: Agricultural Research Station, 1 December 1973 (Halsted 1974). *Milvus m. parasitus* is a summer visitor, seen from late September to early February, common at least in the Maseru-Roma area. Seems uncommon in other parts of the country. Current breeding records needed. Might have decreased in numbers as Jacot-Guillarmod (1963) writes, 'Widely distributed summer visitor in both lowlands and mountains. Nesting in the territory'.

Blackshouldered Kite
Elanus caeruleus
127

Common in the lowland region, but shows some fluctuation during the year. Breeding recorded at the Agricultural Research Station (Davidson, D. *in litt.*). Can also be seen in the lower regions of the mountains (Senqu Valley and upper mountain valley regions): around Qacha's Nek, 7 March 1951 (Rudebeck 1956–63); Marakabei's (Quickelberge 1972); Mphaki, 7 March 1981 (Bonde 1981).

Black Eagle
Aquila verreauxii
131

Uncommon to rare. Can be seen in the higher regions of the mountains in Mokhotlong and Qacha's Nek districts. There have also been sightings from Blue Mountain Pass (Goodfellow MS) and Thaba-Putsoa, 21 January 1973 (Ambrose, D. *in litt.*). Breeding recorded, Upper Quthing Valley (Jilbert, J. *in litt.*).

Tawny Eagle
Aquila rapax
132

'Occasionally seen flying overhead in the lowlands' (Jacot-Guillarmod 1963). Apart from this there are only two sightings: Little Caledon River, October 1966 and Roma, January 1968 (MacLeay 1970). A rare species in Lesotho.

Blackbreasted Snake Eagle
Circaetus gallicus
143

Apart from an old one from Murray (MS), there are three recent records: near St. Michael's, December 1965 (MacLeay 1970); Leshoboro Plateau, 25 February 1981 (Bonde 1981); Tsiquane Woodlot, Leribe district, 20 January 1983 (Davidson, D. *in litt.*). A very rare species in the lowlands of Lesotho.

Bateleur
Terathopius ecaudatus
146

There are two records of this species: Mohale's Hoek, 1956 (Jacot-Guillarmod 1963); near Bushmen's Pass on the Mountain Road, 9 February 1968 (MacLeay 1970). A straggler to Lesotho, very rare.

Palmnut Vulture
Gypohierax angolensis
147

In 1953 one adult and in November 1956 an adult and a juvenile were seen. On the latter date these birds were seen on several occasions (Jacot-Guillarmod 1963). A straggler to Lesotho, very rare.

Steppe Buzzard
Buteo buteo
149

A Palaearctic visitor to Lesotho in the summer months, November to April. Seems common in the lowlands, but rare in the central and higher mountain areas.

Jackal Buzzard
Buteo rufofuscus
152

A resident of the mountains, where it is still common. Specified breeding records needed. Uncommon to rare in the lowlands.

Redbreasted Sparrowhawk
Accipiter rufiventris
155

In the Maseru area it is uncommon. In the rest of the lowlands it is rare, and there are no records from the mountains. Two records of suspected breeding: Seaka Bridge, early November 1973 – probably a pair nesting in poplar copse on the Quthing side of the Senqu River (Brown MS); one pair along Caledon River below U.S. Embassy in Maseru, 26 August 1973. Seen often thereafter along river during spring and summer of 1973. Probably nesting (Halsted 1974). Lesotho is not yet really suitable for this or other *Accipiter* species, but with the growing woodlots this might change.

Little Sparrowhawk
Accipiter minullus
157

First identified 8 March 1974 from Halsted garden (next to the U.S. Embassy), but seen in the area several other times earlier and later (Halsted, D. *in litt.*). Single record.

Black Sparrowhawk
Accipiter melanoleucus
158

Maseru Golf Course, 30 November 1973 (Halsted 1974). Single record.

Little Banded Goshawk
Accipiter badius
159

Three records of this species, all near Roma: a single bird was seen 4 and 5 December 1965, flying and settling in trees in the university village (Goodfellow 1966); one killed weaver in garden (University), January 1967 (MacLeay 1970); at St. John's R.C. Mission near Koro-Koro, 31 December 1978 (Bonde 1981). Very rare in Lesotho.

African Goshawk
Accipiter tachiro
160

There are four records of this species: Mohale's Hoek (Jacot-Guillarmod 1963); University garden in Roma, 3 March 1966 (MacLeay 1970); at Caledon River, 29 March 1973, possibly this species; Agricultural Research Station, 24 August 1975 (Burke MS). Very rare in Lesotho.

Gabar Goshawk
Micronisus gabar
161

There are two records of this species: Qacha's Nek, 9 September 1962 (Jacot-Guillarmod 1963); garden in Roma, 6 January 1966 (MacLeay 1970). Very rare in Lesotho.

Pale Chanting Goshawk
Melierax canorus
162

Two records of this species in the lowlands: Mamathe's (Jacot-Guillarmod 1963); near St. Michael's, 24 February 1967 (MacLeay 1970). There are two doubtful records from Sehlabathebe National Park (Balcomb MS) and an odd mountain record: Mahlabachaneng Pass at 3250 metres, 28 June 1947, one juvenile shot. Bird rose from a francolin carcass. The skin was sent to the British Museum (Vincent, J. *in litt.*). A very rare bird in Lesotho.

African Marsh Harrier
Circus ranivorus
165

Uncommon, if not rare in Lesotho. There are only a few records: one from Sehlabathebe National Park (Balcomb MS); five from the Maseru-Roma area (MacLeay 1970, Bonde 1981, Halsted, D. *in litt.*) and two from Tša-Kholo – one of these is a nesting record (Jilbert, J. *in litt.*). There are not that many localities for this species in Lesotho, apart from Koro-Koro and Tša-Kholo.

Pallid Harrier
Circus macrourus
167

Apart from Jacot-Guillarmod (1963) stating that it is a rather rare summer visitor to the lowlands, there is only one sighting: Koro-Koro, 6 February 1969 (MacLeay 1970). Single record.

Black Harrier
Circus maurus
168

Recorded regularly in Sehlabathebe National Park from November 1978 to April 1979 (Balcomb MS); seen in the Thaba-Tseka area, five individuals (Brown, C. *in litt.*); and three in the Maseru-Roma area (MacLeay 1970, Burke MS, Brown MS). Rare in the lowlands, rare or maybe uncommon in the lower south-eastern parts of the mountains.

Gymnogene
Polyboroides typus
169

Recorded in the Roma area and around some of the bigger plateaux in the lowlands, but not in any numbers (MacLeay 1970, Aspinwall 1973, Ambrose, D. *in litt.*, Halsted, D. *in litt.*). Uncommon if not rare. One sighting from Mokhotlong, 1 December 1973 (Burke MS).

Family   FALCONIDAE: Falcons and kestrels

Seven species reported in Lesotho.

**Peregrine
Falcon**
Falco peregrinus
171

A very rare species with only two recent sightings: flying above St. Peter's Village, January 1965 (MacLeay 1970), and north of Teyateyaneng, 24 November 1974 (Halsted, D. *in litt.*). However Jacot-Guillarmod (1965) writes, 'Nesting Qaoling, September'. This suggests that further careful study is necessary.

Lanner Falcon
Falco biarmicus
172

Common both in the lowlands and in the mountains. Breeding (Jacot-Guillarmod 1963); recent breeding records are however lacking. But it must still be breeding, as it is commonly seen in all parts of the country.

Hobby Falcon
Falco subbuteo
173

So far there have only been four sightings: sewage ponds, Maseru, 9 November 1980 (Bonde 1981); Agricultural Research Station, 11 November 1982; 15 and 22 January 1983 (Davidson, D. *in litt.*).

**Eastern
Redfooted
Kestrel**
Falco amurensis
180

Before 1979 there are only three records: one at Mafeteng 19 January 1914 after tremendous storm – plenty of Lesser Kestrels at the time; several more February 1914 in company with *F. naumanni* (Murray MS); near Quthing, 17 March 1951 (Rudebeck 1956–63).

From 1979 to 1981 there are nine records: old Maputsoe road, 4 February 1979; Roma University, 18 February 1979; on Pitseng road, about 10 km east of Leribe, 11 March 1979; at King's Palace, Maseru, 12 March 1980, 22 March 1980; Mafeteng road, just south of Mazenod, 23 March 1980; at King's Palace, Maseru, 27 December 1980; at Makhalanyane, on Roma road, 8 February 1981; at King's Palace, Maseru, 20 February 1981 (Bonde 1981). Seems more common than the literature indicates, though in small numbers. It is a summer visitor, often in association with *F. naumanni*, and should be sought for where this species is roosting. More material is however needed to determine its actual status.

Rock Kestrel
Falco tinnunculus
181

Common in the whole country, both in the lowlands and in the mountains. Recent breeding records are needed.

Greater Kestrel
Falco rupicoloides
182

There is only one dubious record. In Murray (MS) is written: 'Specimen shot at Mohale's Hoek 2/8/54 (S)'. The year '54' must be a misprint for a later year. The species is only hypothetical.

Lesser Kestrel
Falco naumanni
183

A common and numerous summer visitor, arriving mid-October and leaving late March. It is conspicuous on its roosting places, in the eucalyptus trees of the towns in the lowlands. There are no records from the mountains.

Roosting in thousands every night in tops of bluegums at Mafeteng and Mohale's Hoek (Murray MS).

In recent years large flocks have been reported at King's Palace, January 1973 (Bass, H.G.M. *in litt.*), in 1978–79, 1979–80 and 1980–81, with a maximum of more than 2000 on 20 February 1981 (Bonde 1981). Large flocks have congregated on NUL in the evenings to roost in recent years. By 17 February there were 300–500 roosting daily (Ambrose, D. *in litt.*).

Family

PHASIANIDAE: Francolins, quail, pheasants, partridges, etc.

Six species in Lesotho.

Greywing Francolin
Francolinus africanus
190

The common francolin in the Lesotho lowlands, where it can be found in some of the larger woodlots, e.g. Leshoboro, Tsiquane and Serupane, and in the mountains. Breeding reported from Sanqubetu Valley, 13 December 1961 (Symons, G. *in litt.*).

Redwing Francolin
Francolinus levaillantii
192

Present status uncertain. Jacot-Guillarmod (1963) writes 'Basutoland (McLachlan & Liversidge)'. Reported by G. Maclean (*in litt.*) from Blue Mountain road in 1954.

Orange River Francolin
Francolinus levaillantoides
193

Apart from being collected by Sir Andrew Smith in 1843 near the headwaters of the Caledon River, it has been collected once: just on the border where the Caledon leaves Lesotho near Wepener, 25 May 1922 (Murray MS). Hypothetical.

Swainson's Francolin
Francolinus swainsonii
199

This is a recent addition to the fauna of Lesotho, still with a rather limited distribution. NUL Campus, 9 March 1972 (Ambrose, D. in litt.); Agricultural Research Station, 1969, 1973, 1977 (Brickell 1980); 23 December 1979, 22 March 1980, 23 November 1980, 29 December 1980, 18 January 1981 (Bonde 1981); seen every evening (Davidson, D. *in litt.*). It is said to be spreading in Natal and the Orange Free State, so more records can be expected.

Common Quail
Coturnix coturnix
200

Fairly common in both lowlands and mountains, more often heard than seen. It is said to sing: '*U ka le qeta?*' – 'Can you finish it', for the women, when they are hoeing the fields in December. Breeding suspected but not yet proved.

Harlequin Quail
Coturnix delegorguei
201

Apart from three records at the beginning of the century – one male shot in Leribe in 1904, and one in Masite in 1920 and another one seen in Maseru in 1920 (Murray MS) – there is only one recent record: Rampai's Nek, 1970 (Brickell 1980). Single record.

Family

NUMIDIDAE: Guineafowl

One species in Lesotho.

Helmeted Guineafowl
Numida meleagris
203

This species is fairly common in some areas of the lowlands, e.g. the sandstone gorges around Roma, Maphotong Gorge, Koro-Koro and Raboshabane Gorge (Ambrose, D. *in litt.*). Seen at all times of the year at the Agricultural Research Station, roosts in poplar trees along the Caledon River. Also seen frequently now in some of the larger woodlots (Davidson, D. *in litt.*). Breeding suspected, but not yet proved.

Family   TURNICIDAE: Buttonquails

One species in Lesotho.

Kurrichane **Buttonquail**
Turnix sylvatica
205

At the turn of the century it seems to have been fairly common, as a few were shot every year by Murray, who also caught one about four days old on 29 March 1907 (Murray MS). In recent years there is only one record: grassland in Roma Valley, 26 January 1965 (MacLeay 1970). Very rare.

Family   GRUIDAE: Cranes

Three species reported as rare visitors to Lesotho.

Wattled Crane
Grus carunculata
207

There are seven records from Sehlabathebe National Park: January, April, and December 1976, January 1977 and March 1980 (Balcomb MS); 3–4 January 1976; December 1982 (Coghlan, D. *in litt.*). A rare visitor to the south-eastern corner of the country.

Blue Crane
Anthropoides paradisea
208

There are six records of this species, five of these are from the south-eastern corner of the country. Roma, 23 September 1967 (MacLeay 1970); Letšeng-la-Letsie, early November 1973 (Brown MS); Sehlabathebe National Park, September 1975, October and December 1976 (Balcomb MS) and 3 January 1976 (Cochlan, D. *in litt.*). A rare visitor.

Crowned Crane
Balearica regulorum
209

There are only two records of this species, both from Sehlabathebe National Park, December 1975 and January 1976 (Balcomb MS). Another rare visitor to Lesotho.

Family   RALLIDAE: Rails, crakes, gallinules, moorhens, coots, etc.

Ten species in Lesotho. Only two are common and breeding. Six are hypothetical.

African Rail
Rallus caerulescens
210

There is one not quite satisfactory record of this species, from Jacot-Guillarmod (1963), 'Once seen in Maseru'. Due to its habits it could be a common bird in small numbers, without being noted. Recordings of its voice can help in locating this species. Might be at Koro-Koro and Tša-Kholo. Hypothetical.

Corncrake
Crex crex
211

Apart from two specimens being shot at the beginning of the century: Maseru, 16 February 1907, and Mafeteng, 12 April 1925 (Murray MS), there is only one recent record: Sehlabathebe National Park, 3 January 1976 (Coghlan, D, *in litt.*). Single record.

African Crake
Crex egregia
212

One shot at Maseru, 19 March 1909 (Murray MS). With no recent records it must be regarded as hypothetical.

Spotted Crake
Porzana porzana
214

One shot at the corn exchange near Leribe, 26 June 1912, by M.E. Barrett; specimen in the collections of the Transvaal Museum (Murray MS). No recent records. Hypothetical.

Striped Flufftail
Sarothrura affinis
221

One at Teyateyaneng (Murray MS). Comment by the editor of Murray's notes, J.M. Winterbottom: 'Not a very satisfactory record'. Must be regarded as hypothetical.

Purple Gallinule
Porphyrio porphyrio
223

A few specimens were shot at the beginning of the century (Murray MS). There are five recent records: Tša-Kholo, 1978 (Jilbert, J. *in litt.*); Nye-Nye Dam, 13 May 1979; Leribe Dam, 22 June 1980; Agricultural Research Station, 4 January 1981 (Bonde 1981); Leribe Dam, 3 June 1982 (Davidson, D. *in litt.*). Rare.

Lesser Gallinule
Porphyrula alleni
224

There are three records from the turn of the century (Murray MS), but with no recent material, this species must be regarded as hypothetical.

Moorhen
Gallinula chloropus
226

This species is a common lowland resident of marshes and dams with reed cover. More secretive than *Fulica cristata*, and not seen as often. Not reported from the mountains. Breeding.

Lesser Moorhen
Gallinula angulata
227

There is one not quite satisfactory record of this species from Jacot-Guillarmod (1963), 'Seen once on the edge of Leribe Dam'. Recordings of its voice can help in locating this species. Might be at Koro-Koro and Tša-Kholo. Hypothetical.

Redknobbed Coot
Fulica cristata
228

A common resident and breeding species in the lowlands. There are several breeding records: Agricultural Research Station, 7 January 1974 (Burke MS); eggs collected by J.A. Cottrell, Maseru, 26 January 1922 (James 1970).

Can be seen on all dams which have at least some vegetation. Also recorded from: Sehlabathebe National Park from September to May, except February, (Balcomb MS); Letšeng-la-Letsie, 6 March 1981 (Bonde 1981).

Family

OTIDIDAE: Bustards and korhaans

Five species in Lesotho.

Stanley's Bustard
Neotis denhami
231

There are two sightings from the lowlands: Agricultural Research Station, 25 July 1978 and 19 August 1978 (Bonde 1981); and two from Sehlabathebe National Park, February and March 1979 (Balcomb). A very rare visitor to Lesotho.

Ludwig's Bustard
Neotis ludwigii
232

One shot near Mafeteng, 13 January 1916 (Murray MS). Rare visitor to the lowlands; seen at Maseru, 1956 and recorded from Mamathe's. The latter specimen, a female, was struck by a Lanner (Jacot-Guillarmod 1963). A very rare visitor to the lowlands, if not hypothetical.

Whitebellied Korhaan
Eupodotis cafra
233

Murray (MS) writes, 'A few to be found across the border in O.R.C. and very rarely come into Basutoland', and Jacot-Guillarmod (1963) writes in parenthesis, 'Near Maseru, what was probably this form, was seen on two occasions'. Apart from this there are no definite records of this species. Hypothetical.

Blue Korhaan
Eupodotis caerulescens
234

Fairly common in all lowland districts. Breeding reported: eggs found on several occasions in some of the woodlots (Davidson, D. *in litt.*).

Black Korhaan
Eupodotis afra
239

This is an uncommon, if not rare species, so far only in the lowlands. No breeding records.

Family

ROSTRATULIDAE: Painted snipe

One species in Lesotho. Rare.

Painted Snipe
Rostratula benghalensis
242

There are only two sightings of this species: below Thorn's Store, 17 November 1964 (MacLeay 1970), and Newhauser Dam, Roma Valley, 23 October 1971 (Ambrose, D. *in litt.*). Rare, but due to its skulking, rail-like habits, it could to some extent be overlooked.

Family

CHARADRIIDAE: Plovers

Six species in Lesotho.

Ringed Plover
Charadrius hiaticula
245

A rare Palaearctic visitor to Lesotho; there are only three sightings: Leribe Dam, 7 September 1968, (Dove, R. *in litt.*); Agricultural Research Station, near the farm buildings, August 1973 (Burke MS); New Europe, Maseru, 26 August 1973 (Lexander MS).

Kittlitz's Plover
Charadrius pecuarius
248

There are three sightings of this species, from the same locality or in the vicinity. It has been found breeding at Luma Pan, 17 October 1971 (Ambrose, D. *in litt.*); seen Luma Pan and Tša-Litlama Dam, 7 September 1980 (Bonde 1981). Rare.

Threebanded Plover
Charadrius tricollaris
249

Common in the lowlands from late August to early April, normally seen singly or two together. Apart from one sighting from Leribe Dam, 22 June 1968 (Dove, R. *in litt.*), it seems to leave Lesotho in the winter, so a sighting of about ten at Tša-Litlama Dam, 30 March 1980 (Bonde 1981) could have been birds gathering before leaving Lesotho for the winter. Breeding suspected but not yet proved. There is one sighting from the mountains: river below Malefiloane Clinic, Mokhotlong (about 2650 metres), 2 March 1978 (Bonde 1981).

Grey Plover
Pluvialis squatarola
254

Jacot-Guillarmod (1963) has included this species, but there are no sightings of it from Lesotho so far. Hypothetical.

Crowned Plover
Vanellus coronatus
255

There are two records of this species: St. Michael's, 21 September 1965 (MacLeay 1970); Agricultural Research Station, 15 November 1982 (Davidson, D. *in litt.*). A rare visitor to Lesotho.

Blacksmith Plover
Vanellus armatus
258

Widely distributed in the lowland region throughout the year along dams and even in small marshy places. No records from the mountains. Breeding suspected, but not yet proved.

Family

SCOLOPACIDAE: Sandpipers, snipe, etc.

Ten species in Lesotho. Four of these common.

Turnstone
Arenaria interpres
262

There is a single record of this Palaearctic species: sewage ponds, Maseru, 30 September 1973 (Halsted 1974).

Common Sandpiper
Tringa hypoleucos
264

Common Palaearctic visitor to the lowlands, from mid-September to late March. To be met with along rivers, like the Caledon, on dams such as Nye-Nye, and in marshes like Koro-Koro. One record from the mountains: Khubelu River (3000 metres), seen by G. Maclean (Jacot-Guillarmod 1963).

Wood Sandpiper
Tringa glareola
266

Very common Palaearctic visitor to the lowlands from late August to early April. To be met with on dams like Nye-Nye and Tša-Litlama and in marshes such as Koro-Koro. There is one record from the mountains: 'Just outside Mokhotlong in the high mountains of Basutoland there were two small dams, one surrounded by bushes and rich vegetation, the other one quite open and situated on dry, heavily grazed grassland. In the evening of the 6 April 1951 a single Wood Sandpiper arrived at the latter dam and remained until darkness' (Rudebeck 1956-63).

Marsh Sandpiper
Tringa stagnatilis
269

MacLeay (1970) noted it annually at lake Koro-Koro; since then there has only been one sighting from this locality: 9 September 1978 (Bonde 1981). From September 1978 to February 1981 the water level decreased sharply. There is only one sighting from another part of Lesotho: Morija Dam, 30 September 1979 (Bonde 1981). An uncommon, if not rare Palaearctic visitor to Lesotho.

Greenshank
Tringa nebularia
270

Common Palaearctic visitor to both lowlands and the lower parts of the mountains (Senqu Valley region) in suitable localities. In the collections of the Transvaal Museum there is one skin: Bokong River, a branch of Senqunyane, 7 February 1933, collected by Jacot-Guillarmod. Arriving late August and leaving late March in the lowlands. Has been recorded from Sehlabathebe National Park from October to March (Balcomb MS).

Curlew Sandpiper
Calidris ferruginea
272

Only one record: Koro-Koro, 31 December 1978 (Bonde 1981).

Little Stint
Calidris minuta
274

Uncommon Palaearctic visitor to Lesotho, most often recorded from Koro-Koro in small flocks of about ten (Aspinwall 1973, Bonde 1981, Halsted, D. *in litt.*). Also recorded at Tša-Litlama Dam, 30 March 1980, 4 February 1981; Tša-Kholo, 4 February 1981 (Bonde 1981).

Ruff and Reeve
Philomachus pugnax
284

Common Palaearctic visitor to the lowlands from early September to early April, usually a few seen together, but at favourable places like Koro-Koro as many as 50 may be seen (Jacot-Guillarmod 1963, Aspinwall 1973, Bonde 1981). Seen once at Sehlabathebe National Park, 29 December 1976 (Coghlan, D. *in litt.*).

Ethiopian Snipe
Gallinago nigripennis
286

According to Jacot-Guillarmod (1963) there were at that time no recent records. Since then it has been recorded a number of times: dam six miles from Mazenod on the road to Ha Ramabanta, 21 February 1965 (Goodfellow 1966). Halsted (*in litt.*) mentioned that it has been seen on several occasions in warmer weather in Koro-Koro, from 18 November 1973. Bonde (1981) has four records from Koro-Koro, 31 December 1978, 18 March 1979, 6 January 1980, 25 March 1980; and two records from Leribe Dam, 13 May 1979, 22 June 1980. There are four records from Sehlabathebe National Park, March 1977, January, February and June 1980 (Balcomb MS). Uncommon in the lowlands, rare in the lower parts of the mountains (below 2500 metres).

Curlew
Numenius arquata
289

Apart from a specimen shot on Luma Pan near Mafeteng in 1902 (Murray MS), in recent years there has only been one sighting: Roma, 16 October 1966 (MacLeay 1970). Single record.

Family

RECURVIROSTRIDAE: Avocets and stilts

Two species in Lesotho.

Avocet
Recurvirostra avosetta
294

Rare visitor to Lesotho. One shot by A.T. Bailey near Teyateyaneng, 9 March 1908 (Murray MS). In recent years it has only been recorded at Tša-Kholo: regular visitor in 1976 and 1977 (Jilbert, J. *in litt.*); 27 and 28 June 1977 (Jones, P.J. *in litt.*); 7 January 1979 (Bonde 1981).

Blackwinged Stilt
Himantopus himantopus
295

Apart from a pair shot at Tša-Kholo in 1901 (Murray MS), the first recent sighting is from Koro-Koro, 24 November 1973 (Halsted 1974). Thereafter Bonde (1981) has eleven sightings from July 1978 to January 1981 from both smaller and bigger dams in the lowlands, and Davidson (*in litt.*) has two sightings: Agricultural Research Station, 20 January 1981; Leribe Dam, 23 April 1983. An uncommon, if not rare species, which seems to have increased in numbers in recent years.

Family

BURHINIDAE: Dikkops

One species in Lesotho.

Spotted Dikkop
Burhinus capensis
297

This species is common in the lowland region all year round. A pair nested at Agricultural Research Station, November 1982 (Davidson, D. *in litt.*). There are no records outside the lowland region.

Family

GLAREOLIDAE: Coursers and pratincoles

Three species in Lesotho, all hypothetical.

Burchell's Courser
Cursorius rufus
299

This species has not been noted since 1901 when two large flocks came to Mafeteng (Murray MS). Hypothetical.

Temminck's Courser
Cursorius temminckii
300

Apart from two being shot out of a flock of seven or eight at Maseru, 4 June 1921, and a sighting in Berea District, 1 February 1925 (Murray MS), there are no recent sightings of this species. Hypothetical.

Blackwinged Pratincole
Glareola nordmanni
305

Only mentioned by Murray (MS) as visiting Basutoland rarely and only when locusts are about. No recent records. Hypothetical.

Family

LARIDAE: Skuas, gulls, terns

Three species in Lesotho, all rare.

Greyheaded Gull
Larus cirrocephalus
315

Three records of this species: Letšeng-la-Letsie, December 1970 (Bass, H.G.M. *in litt.*); sewage ponds, Maseru, 3 August 1980 (Bonde 1981); Agricultural Research Station, 2 July 1982 (Davidson, D. *in litt.*). A rare visitor to Lesotho.

Whiskered Tern
Chlidonias hybridus
338

There are four sightings of this species, three of these relating to the same specimens: Agricultural Research Station, 30 December 1973 (Halsted 1974); 28 December 1980, 29 December 1980 (Bonde 1981); Agricultural Research Station, 28 December 1980 (Davidson, D. *in litt.*). Same specimen as Bonde's above. A rare visitor to Lesotho.

Whitewinged Tern
Chlidonias leucopterus
339

There are two sightings of this species: Agricultural Research Station, 18 February 1973 (Lexander MS); Nye-Nye Dam, 4 February 1979 (Bonde 1981). A rare visitor to Lesotho.

Family

PTEROCLIDAE: Sandgrouse

One species in Lesotho, rare.

Namaqua Sandgrouse
Pterocles namaqua
344

Seems to have decreased in numbers; Jacot-Guilllarmod (1963) mentions that 'in former years it was well known in the lowlands, but more recently it is only occasionally heard passing overhead'. There is only one recent record: Agricultural Research Station, sewage ponds (lake 1), 3 June 1982 – a small flock of eight (Davidson, D. *in litt.*).

Family COLUMBIDAE: Pigeons and doves
Five species in Lesotho.

Rock Pigeon
Columba guinea
349

Very common throughout the country and all year. Eggs collected Maseru, 14 February 1923 by J.A. Cottrell (James 1970).

Redeyed Dove
Streptopelia semitorquata
352

The least abundant dove in Lesotho of the genus *Streptopelia*, but it appears to be fairly common. May be seen throughout the year, but more common in summer. So far only recorded from the lowlands. Jacot-Guillarmod (1963) states that it made its first appeared in the lowlands in the early thirties, but it was first heard in Maseru, 5 June 1921. Also located in Leribe, 15 August 1921 and Butha-Buthe 18 August 1921 (Murray MS).

Cape Turtle Dove
Streptopelia capicola
354

A common resident of the lowlands. Eggs collected by J.A. Cottrell, Maseru, 21 December 1921 (James 1970). It is penetrating deeply into the mountains along the river valleys as long as it can find trees. Common in Mokhotlong and two birds were found even higher and several miles further east, in a kraal with five willow trees, December 1947 (Vincent, J. *in litt.*). Makhaleng River near Mountain Road, 11 August 1973 (about 1990 metres); Senqunyane River at Marakabei's, 9 September 1973; Mokhotlong, 1 December 1973 (Burke MS); Qhoali River near Mphaki, 7 March 1981 (Bonde 1981).

Laughing Dove
Streptopelia senegalensis
355

Very common resident of the lowlands. Breeding recorded at Agricultural Research Station (Burke MS). There is one mountain record: Senqunyane River, Marakabei's, 9 September 1973 (Burke MS).

Namaqua Dove
Oena capensis
356

Jacot-Guillarmod (1963) states, 'Widely distributed in the lowlands; nests on the banks of rivers and dongas, either on very low bushes or more often on the ground'. Eggs collected by J.A. Cottrell, Maseru, 17 July 1925 (James 1970). In recent years the number of sightings has decreased sharply, so it seems much rarer than before. Uncommon, if not rare in the lowlands.

Family CUCULIDAE: Cuckoos, coucals, etc.

Six species in Lesotho.

European Cuckoo
Cuculus canorus
374

David Ambrose (*in litt.*) writes, 'This species was added on good authority between the first (1974) and second editions (1976) of the *Guide to Lesotho*. I think the authority was Angela Aspinwall. I am retaining it in the third edition (1984) of the *Guide to Lesotho* as a single record'. Aspinwall (1973) does not mention this species, but knowing David Ambrose I shall follow his judgement and include this species as a single record.

Redchested Cuckoo
Cuculus solitarius
377

Easily identified by its high ringing three-note call, which can be heard from late October to early January. The oldest record is from Leribe, November 1959 by G. Maclean (Jacot-Guillarmod 1963). Summer visitor to Roma (Aspinwall 1973, Ambrose, D. *in litt.*); and Maseru (Burke MS, Halsted, D. *in litt.*, Davidson, D. *in litt.* and Bonde 1981). It seems to have become more common in recent years. As the species prefers to call from a hidden place in high trees it might occur in all the district towns of the lowlands, and in some of the bigger woodlots, if the trees are allowed to grow large enough. A fairly common, but maybe infrequent summer visitor.

Great Spotted Cuckoo
Clamator glandarius
380

There are only some old records of this species: Maseru, 2 March 1909. Also seen at Maseru, 13 November 1918, (Murray MS). Eggs collected by J.A. Cottrell, Mongo, 24 September 1935 (James 1970). If new evidence does not turn up this species must be regarded today as hypothetical.

Jacobin Cuckoo
Clamator jacobinus
382

Apart from a few sightings at the turn of the century (Murray MS), the material gives the following sightings: Hillside in Roma Valley, 5 December 1965, 28 December 1965 (Goodfellow MS); Maphotong Gorge, 3 November 1973; Maseru West, 2 March 1975 (Halsted, D. *in litt.*); Mt. Tsoing, between Maseru and Tša-Kholo, 4 November 1973 – one blackbreasted phase hawking insects (Burke MS); Senqu River, 5 miles north of Quthing, in a patch of natural woodland, early November

1973 – one whitebreasted phase; at Seaka Bridge early November 1973 – one whitebreasted phase (Brown MS); New Europe, Maseru, 19 January 1975 – two pairs, whitebreasted phase (Lexander MS); Sehlabathebe National Park, November 1978 (Balcomb MS). An uncommon, if not rare summer visitor to the lowlands and the lower parts of the mountains.

Diederik Cuckoo
Chrysococcyx caprius
386

Common summer visitor to the lowlands. Its call can be heard from late October onwards. Breeding recorded, with the following hosts: *Passer melanurus* and *Ploceus velatus*. There are two mountain records: Mokhotlong Valley (upper mountain valley region), December 1947 (Vincent, J. *in litt.*); 1 December 1973 (Burke MS).

Burchell's Coucal
Centropus superciliosus
391

Apart from a statement made by MacLeay (1970), 'Seen on several occasions at the pond (NUL), almost annually', there are only two sightings: Mamathe's, May 1943, during a very wet season (Jacot-Guillarmod 1963); Teyateyaneng–Leribe road, 16 April 1972 (Lexander MS). A rare species in Lesotho.

Family

TYTONIDAE: Barn and grass owls

Two species in Lesotho.

Barn Owl
Tyto alba
392

The most common owl in the lowlands according to Jacot-Guillarmod (1963). MacLeay (1970) and Aspinwall (1973) record it as a campus resident at Roma. Difficult to judge whether its status has changed or not, as the material is too thin.

Grass Owl
Tyto capensis
393

Murray (MS) has this species as common. But apart from this there is only the following information: at Leribe, 14 March 1964, nest with four eggs in grass flats near Elliots Dam (Stanford, W. in litt.). Single record. May even be hypothetical today.

Family	STRIGIDAE: Typical owls
	Four species in Lesotho.

Marsh Owl
Asio capensis
395

Recorded from Koro-Koro, 18 November 1973 – at least 25 individuals counted. Thereafter seen in the area in lesser numbers (Halsted, D. *in litt.*); 18 August 1974 (Ambrose, D. *in litt.*); 6 January 1980 (Bonde 1981); Thaba-Khupa Farming Project, 21 December 1978, 25 December 1979, 27 December 1980. A juvenile owl, which could not yet fly, was seen around 1 February 1981, and recorded by Bonde (1981); Masianokeng Woodlot, 16 November 1981; Tsiquane Woodlot, Leribe, 14 December 1981; Agricultural Research Station, 28 October 1982 (Davidson, D. *in litt.*). This owl seems fairly common in suitable, marshy areas in the lowlands. Breeding.

Scops Owl
Otus senegalensis
396

This species has been seen only twice: Maseru, 4 January 1907 (Murray MS); and July 1964 (Stanford, W. *in litt.*). A very rare visitor to Lesotho.

Cape Eagle Owl
Bubo capensis
400

In the collections of the Transvaal Museum is a skeleton of this species from Leribe, 26 September 1977, collected by R. Dean. Single record.

Spotted Eagle Owl
Bubo africanus
401

Jacot-Guillarmod (1963) writes, 'General in the lowlands, but not often seen'. With only five sightings in the lowlands in recent years, together with seven from Sehlabathebe National Park it is difficult to pass any judgement on the status of this species.

Family	CAPRIMULGIDAE: Nightjars
	Two species in Lesotho.

European Nightjar
Caprimulgus europaeus
404

Uncommon Palaearctic visitor to Lesotho. There is not much information. Jacot-Guillarmod (1963) writes, 'Fairly common summer visitor to the lowlands', and MacLeay (1970) writes, 'Occasionally seen flying at dusk in summer'. Apart from this, there is one record: Maseru Golf Course, 19 January 1976, and several evenings thereafter (Lexander MS).

Mozambique Nightjar
Caprimulgus fossii
409

There is one record of this species: Agricultural Research Station, mid-October to mid-December 1973. On some occasions the bird was also seen flying and on 6 January 1974 the bird was still calling (Burke MS). Single record.

Family

APODIDAE: Swifts

Six species in Lesotho.

European Swift
Apus apus
411

It is very difficult to distinguish between the European Swift and the African Black Swift in the field There are only five sources which mention this species: Jacot-Guillarmod (1963) writes in parenthesis, 'Presumably occurs together with the next species'; MacLeay (1970) writes, 'Summer migrant, probably frequent, usually in flocks'; Aspinwall (1973) '*A.apus/A.barbatus*: these two species are so similar in flight that it is best to describe them together. Both are seen over Roma Valley and the Campus, but the European Swift is a summer migrant, whereas the Black Swift is a resident of Africa, though seen in the valley chiefly in the summertime'. C. Brown (*in litt.*) 'escarpment, near Sani Pass Border Post'; Bonde (1981) writes, 'I have not felt it possible safely to separate these two species, though I am most inclined to make all swifts seen *A. barbatus*. On several occasions I have noted that the secondaries were very light seen from below, contrasting with the rest of the wing, which pulls in the direction of *A. barbatus*'. Status uncertain.

Black Swift
Apus barbatus
412

This species is a common visitor to Lesotho. Breeding recorded at junction of Lekhalabaletsi and Sanqubetu Rivers, 3 December 1947: 'By means of a rope reached a nesting colony of about 20 pairs, all nests contained feathered juveniles' (Vincent, J. *in litt.*).

Whiterumped Swift
Apus caffer
415

Summer visitor to Lesotho, but seems less numerous than the two other 'white-rumped swifts' at least in the Maseru area. Breeding in Drakensberg using nest of *Hirundo cucullata* (Symons 1919). Eggs collected by J. A. Cottrell, Maseru, colony on Caledon River, 14 February 1923 (James 1970).

Horus Swift
Apus horus
416

Common in the lowlands, can be seen from mid-October to late March. Also found in the mountains, at least recorded from the Mokhotlong district. Breeding recorded: Upper Lekhalabaletsi River, 1 December 1947 (Vincent, J. *in litt.*); on trip from Giants Castle to Sanqubetu Valley, 10 December 1960, found nesting in a hole in a bank, (Symons, G. *in litt.*); Bushmen's Pass, late summer 1974, two juveniles (Halsted, D. *in litt.*).

Little Swift
Apus affinis
417

This swift might be a newcomer to Lesotho as it has only been noted a few times before the end of the 1970s. Bonde (1981) writes, 'Have seen it commonly in Maseru, first recorded 21 August 1978. Normally November to March. Often gathering in flocks, screaming, and flying wildly, chasing each other, very much like *Apus apus* in Europe. Also recorded: Koro-Koro, 31 December 1978; Tša-Kholo, 7 January 1979; Roma, 18 February 1979. Breeding recorded: on the north end of Majara House, Agricultural College, Maseru, 28 March 1981 – a breeding colony of about 10 nests'.

Alpine Swift
Apus melba
418

Common summer visitor both to the lowlands and the mountains. Can be seen from mid-August to late March. Often seen in small numbers together with other swifts.

Family

COLIIDAE: Mousebirds

Three species in Lesotho.

Speckled Mousebird
Colius striatus
424

This seems to be another newcomer to Lesotho, so far found only in some parts of the lowlands. Jacot-Guillarmod (1963) has only noted it from the extreme southern portion of Lesotho. In the 1960s it established itself in Maseru and around 1970 it had become common (Stanford, W. *in litt.*; Halsted, D. *in litt.*). An advance party of four birds arrived on Roma Campus in late 1971. Since then they have become a pest in the gardens; also occurs in indigenous bush in the Roma vicinity (Ambrose, D. *in litt.*). Also observed in Leshoboro Woodlot north of Maseru, 25 August 1980, 15 October 1980, 15 February 1981 (Bonde 1981).

Whitebacked Mousebird
Colius colius
425

Leribe, in the garden of R. Dove in July 1970. Seen for the first time since arrival in 1952. Single record.

Redfaced Mousebird
Urocolius indicus
426

Uncommon winter visitor to the lowlands, so far all recent records from Maseru (Halsted 1974, Bonde 1981, Burke MS). It is mentioned by Murray (MS), 'First seen Mafeteng, 1 June 1915, left some time in October 1915'. It is not mentioned in Jacot-Guillarmod (1963).

Family

TROGONIDAE: Trogons

One species in Lesotho, hypothetical.

Narina Trogon
Apaloderma narina
427

There is only one old record of this species: Moqeni, 21 July 1923 (Murray MS). Today, hypothetical.

Family

ALCEDINIDAE (HALCYONIDAE): Kingfishers

Three species in Lesotho, common.

Pied Kingfisher
Ceryle rudis
428

This species has commonly been seen at the Agricultural Research Station, Maseru from October to May (Bonde 1981, Davidson, D. *in litt.*). Rare in the rest of the lowlands. No records from the mountains.

Giant Kingfisher
Megaceryle maxima
429

Common in the lowlands throughout the year. It nests in holes in the bank of the Caledon River, Maseru area (Burke MS). Seen Marakabei's, 2 July 1970, (Ambrose, D. *in litt.*); and seven times at Sehlabathebe National Park (Balcomb MS).

Malachite Kingfisher
Alcedo cristata
431

Widespread throughout the year in the lowlands, but not seen very often. There is one mountain record: Oxbow (+2500 metres), early November 1973 (Brown MS).

Family

MEROPIDAE: Bee-eaters

One species in Lesotho.

European Bee-eater
Merops apiaster
438

Jacot-Guillarmod (1963) writes, 'Very occasionally flies over from the Orange Free State into the lowlands'. With no other information must be regarded as hypothetical.

Family

CORACIIDAE: Rollers

Two species in Lesotho, rare and uncommon.

European Roller
Coracias garrulus
446

Uncommon, if not rare, and irregular summer visitor to the lowlands. One sighting from the mountains: Mokhotlong, 8–10 December 1962 (Stanford, W. *in litt.*).

Lilacbreasted Roller
Coracias caudata
447

Rare and irregular summer visitor to the lowlands.

Family	UPUPIDAE: Hoopoe
	One species in Lesotho.
Hoopoe *Upupa epops* 451	Common in the lowlands, at least in the Maseru-Roma area. There are no sightings from the mountains. Breeding: Agricultural Research Station, 1974 (Burke MS); Old Europa, Maseru, 1979–80 and 1980–81 (Bonde 1981).
Family	CAPITONIDAE: Barbets
	Three species in Lesotho.
Blackcollared Barbet *Lybius torquatus* 464	There are three records of this species: Maseru, 1960 (Jacot-Guillarmod 1963); Khotsong Gardens, 1970 (Brickell 1980); 15 miles from Maseru on the North Road, 3 September 1973 (Halsted 1974). Rare.
Pied Barbet *Tricholaema leucomelas* 465	Uncommon visitor to the Maseru area and there are no recent records from other parts of the lowlands.
Redfronted Tinker Barbet *Pogoniulus pusillus* 469	There are a few records of this species from Maseru: June every year, but not common, (Stanford, W. *in litt.*); New Europe, Maseru, 11 August 1973 (Lexander MS); Maseru, 1 June 1983, seen by Keith Richardson, (Davidson, D. *in litt.*).

Family INDICATORIDAE: Honeyguides

Three species in Lesotho.

Greater Honeyguide
Indicator indicator
474

Stanford (*in litt.*) recorded the Greater Honeyguide in Maseru from callsite at the junction of the golf course stream and the Caledon River from May 1962 to December 1969, as well as in his garden where it parasitized a Hoopoe in November 1963. Also Maseru, late May 1975 – single female, photographed in garden (Halsted, D. *in litt.*). Present status is uncertain.

Lesser Honeyguide
Indicator minor
476

There are three sightings of this species: Maseru, Garden of Remembrance, 3–7 November 1963 (Stanford, W. *in litt.*); along the Caledon, next to garden, August and September 1970 (Bass, H.G.M. *in litt.*); Caledon River below the polo ground, 12 August 1978 (Bonde 1981). Rare.

Sharpbilled Honeyguide
Prodotiscus regulus
478

Seen in Roma Valley in summer in wattles along stream banks. Rather rare (Aspinwall 1973); Maseru, in garden, 26 November, 7 and 21 December 1963 (Stanford, W. *in litt.*); golf course, Maseru, 30 November 1973 (Halsted 1974). Rare.

Family PICIDAE: Woodpeckers

Two species in Lesotho.

Ground Woodpecker
Geocolaptes olivaceus
480

A common bird in Lesotho up to at least 3000 metres. In the lowlands easily met with on rocky hillsides, as at Morija, or in old dongas with some vegetation, as at Ha Khechane Woodlot. Also in places with indigenous bush still intact, like the sides of the Leshoboro Plateau, i.e. areas not under cultivation.

Cardinal Woodpecker
Dendropicos fuscescens
486

This species has been reported in the gardens of Maseru and Roma and in the trees along the Caledon River. Uncommon, if not rare. Further study needed to determine its actual status in the lowlands.

Family		JYNGIDAE: Wrynecks
		One species in Lesotho.

Redthroated Wryneck
Jynx ruficollis
489

There is only one record of this species: Caledon River below Maseru, December 1962 (Jacot-Guillarmod 1963). Single record.

Family　　　ALAUDIDAE: Larks

Ten species reported from Lesotho.

Rufousnaped Lark
Mirafra africana
494

Recorded from the Maseru-Roma area of the lowlands (MacLeay 1970, Aspinwall 1973, Bonde 1981). Davidson (*in litt.*) has two records from outside this area: below Leshoboro Plateau, 6 December 1979; Kolonyama, 12 November 1980. Uncommon.

Clapper Lark
Mirafra apiata
495

Jacot-Guillarmod (1963) writes, 'Widely distributed and fairly common in the lowlands', and both MacLeay (1970) and Aspinwall (1973) repeat this for the Roma area. As this species is one of the more difficult to identify, if its song is not known, it might have been overlooked, or rather seen but not recognized by other birdwatchers.

Rudd's Lark
Mirafra ruddi
499

There is only one record of this species: Oxbow, February 1961, seen by G. Maclean (Jacot-Guillarmod 1963). The following statement in Clancey (1980) 'East Griqualand . . . O.F.S., lowland Lesotho . . .' must in view of this be queried. So far only seen once from the high mountains of the northern part of Lesotho. Single record.

Longbilled Lark
Mirafra curvirostris
500

According to Jacot-Guillarmod (1963), 'Widely distributed in the lowlands, but not common'. In recent years there are only four sightings: near the Little Caledon River, October 1965, seen once (MacLeay 1970); Thaba-Bosiu, September 1971;

Liphiring, January 1973, (Bass, H.G.M. *in litt.*); next to the flour mill, Maseru, 26 December 1978 (Bonde 1981). It has also been seen twice at Sehlabathebe National Park, January and July 1979 (Balcomb MS). Uncommon, if not rare in Lesotho.

Spikeheeled Lark
Chersomanes albofasciata
506

Suspected from the south-western part of the country, but only one sighting: about 25 km north of Mafeteng on the Maseru–Mafeteng road, 23 January 1980 (Williamson, J. *in litt.*). Single record.

Redcapped Lark
Calandrella cinerea
507

Common both in the lowlands and the mountains, where it can be seen even in the highest parts. Breeding (Quickelberge 1972).

Pinkbilled Lark
Spizocorys conirostris
508

Recorded twice: Roma Valley, the lower slopes of the north wall, 4 April 1965 (Goodfellow 1966); on field near Luma Pan, north of Mafeteng, 30 March 1980 (Bonde 1981). Rare.

Thickbilled Lark
Galerida magnirostris
512

Common in the mountain and high mountain regions. Not found below 2500 metres. The race *montivaga* described from the confluence of the Lekhalabaletsi and Sanqubetu Rivers by Vincent (1948). Breeding recorded by Vincent (*in litt.*) and Quickelberge (1972). Once seen in the Mafeteng district, October 1970 (Bass, H.G.M. *in litt.*).

Chestnutbacked Finchlark
Eremopterix leucotis
515

Jacot-Guillarmod (1963) writes, 'A winter visitor to the lowlands, where it breeds', and MacLeay (1970) writes, 'Not uncommon during winter in cultivation and grassland'. Thereafter there are only two definite records: Ha Toloana, 1970 (Brickell 1980); and on fields near the Little Caledon River, August 1970, (Bass, H.G.M. *in litt.*). It seems to have declined so it must now be regarded as rare. Records of breeding needed.

Greybacked Finchlark
Eremopterix verticalis
516

Of this species there are only two records: Near Maseru, May 1955, seen by J. Winterbottom (Jacot-Guillarmod 1963); Tseka near Maseru, 1970 (Brickell 1980). Rare.

Family

HIRUNDINIDAE: Swallows and martins

Eleven species in Lesotho.

European Swallow
Hirundo rustica
518

A Palaearctic visitor to Lesotho, which can be seen from September to early April and is common in the lowlands. It can also be seen in small numbers in the mountains up to 2100 metres. Generally it is not as common as *Hirundo albigularis* and *H. cucullata*.

Whitethroated Swallow
Hirundo albigularis
520

Common in Lesotho, both in the lowlands and in the mountains up to 3000 metres (mountain region). It is not recorded in the high mountain region. It can be seen from September to April, but seems to arrive about two weeks earlier than *Hirundo cucullata* (Bonde 1981). Eggs collected by J.A. Cottrell, Makhaleng, 3 January 1926 (James 1970). Breeding: Sanqubetu Valley, 14 January 1967 (Symons, G. in litt.).

Pearlbreasted Swallow
Hirundo dimidiata
523

Jacot-Guillarmod (1963) writes, 'A fairly common summer visitor to the lowlands'. Since then it seems to have decreased as MacLeay (1970) writes, 'Occasionally seen in summer, November 1964, not common'. There is only one specified record: Little Caledon, August 1970 (Bass, H.G.M. *in litt.*). It must be regarded as a rare summer visitor to the lowlands of Lesotho.

Greater Striped Swallow
Hirundo cucullata
526

This species is common in Lesotho, both in the lowlands and in the mountains up to 3200 metres. Can be seen from September to April. Breeding: Roma (MacLeay 1970, Aspinwall 1973); Lekhalabaletsi Valley (Vincent, J. *in litt.*). Eggs collected by J.A. Cottrell, Maseru, 13 February 1923, 26 December 1924 (James 1970).

Lesser Striped Swallow
Hirundo abyssinica
527

There is apparently only one breeding record of this species: Maseru, 14 January 1918, collected by J.A. Cottrell (James 1970). The National Museum of Zimbabwe was contacted and replied: 'White swallow eggs are difficult to assess, but I have reservations about this clutch which looks a bit glossy, with dark marks at the thick pole, which may only be nest stains. The measurements are: 20.0 x 14.3, 21.5 x 14.4, 19.7 x 14.3, 20.7 x 14.5. On the card to the eggs is written: "Bird shot and identified". There is however no trace of this skin'. Hypothetical.

South African Cliff Swallow
Hirundo spilodera
528

There are two records of this species: Leribe (no date) (Jacot-Guillarmod 1963) and Mafeteng, 20 November 1971, (Ambrose, D. *in litt.*). Rare. It is however breeding in the Orange Free State close to the Lesotho border, so it might stray into the lowlands in small numbers.

Rock Martin
Hirundo fuligula
529

Common and resident throughout the year, but may be less numerous in winter. Breeding: Roma (Aspinwall 1973). Eggs collected by J.A. Cottrell, Maseru, 4 January 1920 (James 1970); and by G. Symons, Sanqubetu Valley, 10 December 1960 (Symons, G. *in litt.*).

House Martin
Delichon urbica
530

Rare Palaearctic summer visitor to the lowlands, with about five sightings: water reservoir near Leribe, 30 March 1951 (Rudebeck 1956–63); Malubalube Stream between Teyateyaneng and Mamathe's, 4 February 1958 (Jacot-Guillarmod 1963); Koro-Koro, 6 January 1980; Maseru, next to the Sheepskin Factory, 4 January to 15 February 1981 (Bonde 1981).

Sand Martin
Riparia riparia
532

Palaearctic migrant which has only been recorded once: Koro-Koro, 31 December 1978 (Bonde 1981). Single record.

Brownthroated Martin
Riparia paludicola
533

Common resident throughout the year, less numerous in winter. Both in the lowlands and in the mountains up to 3200 metres (high mountain region). Breeding: Motete River, December 1959 (G. Maclean in Jacot-Guillarmod 1963). Eggs collected by J.A. Cottrell, Maseru, 14 February 1923 (James 1970).

Banded Martin
Riparia cincta
534

Rare to uncommon summer visitor to Lesotho. Rare in the lowlands, and uncommon in the Sehlabathebe National Park area.

Family

DICRURIDAE: Drongos

One species in Lesotho.

Forktailed Drongo
Dicrurus adsimilis
541

Apart from two at the beginning of the century (Murray MS), there are no records of this species. Hypothetical.

Family

ORIOLIDAE: Old-world orioles

Two species in Lesotho.

European Oriole
Oriolus oriolus
543

Apart from Jacot-Guillarmod (1963), who writes, 'A rare migrant to the lowlands, only two specimens having been seen in thirty years, one of them in juvenile plumage', there is one record: Maseru, December 1972 (Bass, H.G.M. *in litt.*). Rare.

Blackheaded Oriole
Oriolus larvatus
545

Rare. Only two sightings: Maseru, 24 September 1922 (Murray MS); Maseru, 17 July 1964, seen by Stanford, (Murray 1964).

Family

CORVIDAE, Crows, jays, etc.

Three species in Lesotho.

Black Crow
Corvus capensis
547

Common, but not as numerous as other Corvidae, normally seen in pairs or singly. Breeding: Sanqubetu Valley, at 2600 metres, October 1946 (Vincent, J. *in litt.*).

Pied Crow
Corvus albus
548

This species is confined to the lowlands where it is common. It seems, however, to avoid the densely populated areas like Maseru. Eggs collected, Mongo, 24 September 1935 (James 1970).

Whitenecked Raven
Corvus albicollis
550

Very common both in the lowlands and in the mountains even in the highest parts. Breeding: near Mokhotlong at 2600 metres, October 1946 – pair seen at nest; in December 1947 – adults seen feeding a full grown juvenile (Vincent, J. *in litt.*); Lancer's Gap, 15 September 1973 – incubating in a newly built nest; 5 October 1973 – one immature (Burke MS).

Family

PARIDAE: Tits

Two species reported from Lesotho; one of these, however, is hypothetical.

Southern Grey Tit
Parus afer
551

This is an uncommon species in the mountains to be met with where it can find trees and scrub, from Sani Top in the east to Marakabei's in the west (Jacot-Guillarmod 1963, Quickelberge 1972).

Ashy Tit
Parus cinerascens
552

In Clancey (1980) this former subspecies of the Grey Tit *Parus afer* has been made a separate species. The race *P.c. orphnus* should be distributed: 'lowlands of northern Lesotho, . . .' In the collected material there is no evidence, however, that this is true. It is, however, always a problem, when a new species is nominated, to discriminate between the records and divide them. There is no 'Grey Tit material', which clearly belongs to this new species. The most westerly record of 'Grey Tits' on subspecies level are from Marakabei's and this is of the subspecies *Parus afer arens* (Quickelberge 1972). In the collections of the Transvaal Museum is a *Parus afer* skin from Soosa, Maluti Mountains, collected by Jacot-Guillarmod in 1933, which has not been examined. Hypothetical.

Family	PYCNONOTIDAE: Bulbuls
	One species in Lesotho.

Redeyed Bulbul
Pycnonotus nigricans
567

Common throughout the year in the lowlands, preferring gardens and areas with indigenous bush. There are some records from the lower parts of the mountains up to 1950 metres, which might indicate that it follows the rivers into the mountains in the Quthing district to some extent (Quickelberge 1972, Bonde 1981).

Family	TURDIDAE: Thrushes, chats, robins, etc.
	Fifteen species in Lesotho.

Kurrichane Thrush
Turdus libonyana
576

David Ambrose (*in litt.*) writes, 'This species was added on good authority between the first (1974) and second editions (1976) of the *Guide to Lesotho*. I think the authority was Angela Aspinwall. I am retaining it in the third edition (1984) of the *Guide to Lesotho* as a single record'. Aspinwall (1973) does not mention this species, but knowing David Ambrose, I shall follow his judgement and include this species as a single record.

Olive Thrush
Turdus olivaceus
577

Common in the Maseru area, often seen along the Caledon. Might be found in the gardens of the District Headquarters, and along the rivers with indigenous bush, in the lowlands. Further material needed.

Cape Rock Thrush
Monticola rupestris
581

Found in the lowlands and foothills on dry rocky slopes and in the lower regions of the mountains up to 2500 metres (upper mountain valley). Fairly common. Not found in the high north and north-western parts of Leribe, Butha-Buthe, Tsaba-Tseka and Mokhotlong districts. It appears to share the mountains with *Monticola explorator* according to altitude. It inhabits the lower areas, while *M. explorator* takes the higher ones. Must to some extent follow the rivers into the mountains, as it has been seen at Koma-Koma Bridge, 3 August 1987 and 4 June 1979 (Bonde 1981).

Sentinel Rock Thrush
Monticola explorator
582

Common in the mountains from the foothill region up to the high mountain region. Breeding: between Blue Mountain Pass and Marakabei's, 15 to 19 September 1968 (Quickelberge 1972). Eggs collected by J.A. Cottrell, Malutsonyane Falls, 1 January 1926; Thaba Putsna, 4 January 1926 (James 1970).

Mountain Chat
Oenanthe monticola
586

Common throughout the year on dry rocky slopes in the lowlands, foothills, and in the lower regions of the mountains up to about 2500 metres. Its distribution in the mountains clearly follows the riverbeds. It can penetrate as far as Mokhotlong and still be below 2500 metres. Breeding: Mokhotlong (Symons 1919). Six clutches collected by J.A. Cottrell, Malutsenyane Falls, December 19??; Maseru, 11 December 1923; 18 January 1925 (James 1970).

Capped Wheatear
Oenanthe pileata
587

Jacot-Guillarmod (1963) has written in parenthesis, 'Placed in the list on hearsay only' and Clancey (1980) writes, 'Cape south of Orange . . . Lesotho lowlands'. There is however no positive information in the collected material, so it must be regarded as hypothetical.

Buffstreaked Chat
Oenanthe bifasciata
588

Only one record of this species: eggs collected by J.A. Cottrell, Quthing, 11 October 1929 (James 1970). A small population might have occurred in the Quthing district, and could still be there, unrecorded, as the ornithological activity has been low in this part of Lesotho. Single record.

Familiar Chat
Cercomela familiaris
589

Common resident in the lowlands, where it can be found on rocky hillsides and in dongas. Penetrates the mountains to some extent following the riverbeds, but so far not recorded above 2300 metres (upper mountain valley region). It appears to share the mountains with *Cercomela sinuata* according to altitude, inhabiting the lower areas, *C. sinuata* the higher ones. Breeding (Quickelberge 1972). Eggs collected by J.A. Cottrell, Quthing, 10 October 1929 (James 1972).

Sicklewinged Chat
Cercomela sinuata
591

Common on high ground, but not so far recorded below 2200 metres. Breeding (Quickelberge 1972). Eggs collected by R.E. Symons, Sanqubetu Valley, 15 November 1915 (Transvaal Museum *in litt.*); G. Symons, Sanqubetu Valley, 10 December 1960 (*in litt.*).

Mocking Chat
Thamnolaea cinnamomeiventris
593

Seen once on cliffs north of Roma, December 1966 (MacLeay 1970). There is also a dubious record of eggs collected by J.A. Cottrell, Quthing, 14 October 1929 (James 1970). The data card in the Museum reads as follows: 'Mocking Chat. Collected by African herdboy on 14 October 1929 at Quthing, Basutoland, Southern Africa. No. of eggs in set, 3. Set mark C1138. Identity uncertain. Incubation fresh. From the description of the birds and nest in old swallow's nest, I concluded that this set of eggs could be of the species named above. Identification very tentative. J.A. Cottrell'.

Comment from Mrs. P. Lorber, National Museum of Zimbabwe: 'The clutch itself looks out of place in the Mocking Chat drawer of bluish-white eggs streaked with brown, because the eggs are cream with only the faintest of buffish-brown streaks, and rather highly glossed. The situation could very well be Mocking Chat, as Cottrell suspected, but I would be very cautious of including this clutch if you have any other grounds for believing it to be incorrect. They are equally incorrect for *Oenanthe bifasciata* and *Oenanthe monticola* and one is rather at a loss to know what else they could be. The measurements: 24.8 x 18.7 ; 25.1 x 18.9 ; 24.4 x 19.0'. Had it not been recorded from Roma, its status would be hypothetical. Now it is a single record.

Anteating Chat
Myrmecocichla formicivora
595

Rare, but can be seen in small numbers along the border with the Orange Free State. There are two recent records: Leshoboro Plateau, 24 August 1980 (Bonde 1981); Tsiquane Woodlot, Leribe, 5 July 1982 (Davidson, D. *in litt.*).

Stonechat
Saxicola torquata
596

Very common in the mountains, uncommon in the lowlands, where it is mostly a winter visitor. Breeding: Sanqubetu Valley, 13 December 1961, 14 December 1964, 11 December 1979 (Symons, G. *in litt.*). Eggs collected by J.A. Cottrell, Thaba Putsna, 4 January 1926 (James 1970).

Cape Robin
Cossypha caffra
601

Common in the lowlands in gardens and in indigenous bush. In the mountains it can be seen up to 2500 metres (upper mountain valley region) wherever there is suitable habitat. Suspected breeding: Moletsana, 11 September 1968 (Quickelberge 1972).

Orangebreasted Rockjumper
Chaetops aurantius
612

Common in the mountains, especially above 2000 metres. Not reported from the lowland region and foothills. Eggs collected by G. Symons, Sanqubetu Valley, 10 December 1961 (Symons, G. *in litt.*).

Karoo Robin
Erythropygia coryphaeus
614

Uncommon, if not rare in the lowlands, to be met with on rocky hillsides with indigenous bush. There are two recent records: behind White City, Maseru, 17 August 1980; Leshoboro Plateau, 25 February 1981 (Bonde 1981).

Family

SYLVIIDAE: Warblers, cisticolas, prinias, etc.

Twenty species in Lesotho. Warblers present an identification problem, because of their small size, dull colours, and often skulking habits, and only in recent years have recordings of their voices become available. Material for some of the species is, therefore, rather thin, making it difficult to assess the actual status of a certain species.

Whitethroat
Sylvia communis
620

David Ambrose (*in litt.*) writes, 'This species was included in the first edition (1974) of the *Guide to Lesotho*, because of a definite record from Angela Aspinwall'. Aspinwall (1973) does not mention this species, but knowing David Ambrose, I shall follow his judgement and enter this species as a single record.

Titbabbler
Parisoma subcaeruleum
621

In Clancey (1980) is written 'Natal interior . . . Lesotho lowlands . . .' This must however be queried, as there is only one record of this species: Maseru, May 1955 (Jacot-Guillarmod 1963). Single record.

Layard's Titbabbler
Parisoma layardi
622

Fairly common in the higher regions of the mountains, where it can find cover, down to the area around Molimo Nthuse. Not found in foothills and lowland region. Breeding: Sanqubetu Valley, 13 December 1961 (Symons, G. *in litt.*).

African Marsh Warbler
Acrocephalus baeticatus
631

Common in the lowlands, where it can be heard from even small patches of reeds and trees near or in water. Can be heard from late October to early February. Recorded by G. Maclean from Malibamatso River below 2600 metres, November 1959 (Jacot-Guillarmod 1963), so it might to some extent penetrate the mountains along the river beds. So far no winter records. Breeding: Agricultural Research Station, 7 January 1974 (Burke MS).

Cape Reed Warbler
Acrocephalus gracilirostris
635

Recorded six times: Agricultural Research Station, 1 December 1973 (Halsted 1974); Thaba-Khupa Farm Project, 22 October 1979; Agricultural Research Station, 28 December 1980, 18 January 1981; Leribe Dam, 25 January 1981; Morija Dam, 29 January 1981 (Bonde 1981). Is this a newcomer to Lesotho, or is the lack of records due to birdwatchers not having known its voice? Common in a small number in suitable habitats.

Barratt's Warbler
Bradypterus barratti
639

Jacot-Guillarmod (1963) writes, 'Found along the eastern border'. Apart from this there are no records of this species, but it is reported in Natal in the squares bordering Lesotho (Cyrus 1980), so it might be possible in the south-eastern area of Lesotho, Qacha's Nek and Sehlabathebe. Hypothetical.

Willow Warbler
Phylloscopus trochilus
643

Uncommon Palaearctic visitor in summer to the lowlands. Normally associated with fairly high trees. Recorded from late October to mid-March. At the end of its stay its pleasant little song can be heard.

Barthroated Apalis
Apalis thoracica
645

David Ambrose (*in litt.*) writes, 'This species was added in the second edition (1976) of the *Guide to Lesotho*, because of a definite record from Angela Aspinwall'. Aspinwall (1973) does not mention this species, but knowing David Ambrose I shall follow his judgement and include this species as a single record.

Longbilled Crombec
Sylvietta rufescens
651

There are three sightings of this species: Maseru (Jacot-Guillarmod 1963); Masitise Mission, Quthing, 8 September 1967 (Quickelberge 1972); Maseru, in the trees along the Caledon River, 28 August 1970 (Bass, H.G.M. *in litt.*). Rare.

Grassbird
Sphenoeacus afer
661

So far only recorded in the Maseru-Roma area, where it can be seen on hillsides with dense indigenous bush. Uncommon, if not rare here. But visits to its habitat might reveal a bigger population. Breeding: on the slopes on the north wall of the Roma Valley, 16 January 1965 (Goodfellow 1965).

Fantailed Cisticola
Cisticola juncidis
664

According to Jacot-Guillarmod (1963) this species is 'generally distributed throughout the lowlands', and MacLeay (1970) writes 'frequent in grassland especially near dongas'. Apart from this there is only one recent record: Koro-Koro, 31 December 1978 (Bonde 1981). Must still be common in the lowlands.

Desert Cisticola
Cisticola aridula
665

According to Jacot-Guillarmod (1963) this species is 'generally distributed throughout the lowlands', and Aspinwall (1973) writes 'sometimes seen amongst the weedy area to the left of the road that runs from the entrance to the campus past the Vice-Chancellor's lodge'. Apart from this it has been recorded four times on the Agricultural Research Station; 28 January 1979, 23 November 1980, 28 December 1980, 18 January 1981 (Bonde 1981). Must still be common in the lowlands.

Cloud Cisticola
Cisticola textrix
666

According to Jacot-Guillarmod (1963) this species is 'generally distributed in the lowlands'. Apart from this there are four records: Koro-Koro, 18 March 1979, 18 November 1979, 6 January 1980; fields near Luma Pan, 30 March 1980 (Bonde 1981). Must still be common in the lowlands.

Ayres' Cisticola
Cisticola ayresii
667

According to Jacot-Guillarmod (1963) this species is 'generally distributed both in the lowlands and in the Maluti. Common on Malibamatso River below 2600 metres'. Occurs only very sparsely at the higher levels. One seen and heard in the

Lekhalabaletsi and another in the Sanqubetu Valley, but the birds become a great deal more frequent at the Mokhotlong level (Vincent, J. *in litt.*). Eggs collected by G. Symons, Sanqubetu Valley, December 1971 (*in litt.*).

Wailing Cisticola
Cisticola lais
670

Common in the mountains from upper mountain valley region into the high mountain region, where it can be found right at the top at least in summer. There is one lowland sighting: Leshoboro Plateau, 25 February 1981 (Bonde 1981). Might occur in the lowlands on rocky hillsides with indigenous bush.

Levaillant's Cisticola
Cisticola tinniens
677

Common in the lowlands throughout the year, associated with reeds and bushes over water. Common, but less numerous in the mountains following the riverbeds.

Neddicky
Cisticola fulvicapilla
681

Common in the lowlands throughout the year. According to Quickelberge (1972) not seen beyond the Maluti escarpment. Breeding: Hlotse, garden of R. Dove, December 1968 (Dove, R. *in litt.*).

Tawnyflanked Prinia
Prinia subflava
683

Two sightings of this species: at Ha Khotso rock paintings, 6 July 1973 (Burke MS); Holiday Inn, Maseru, 9 August 1973 (Halsted 1974). Rare.

Blackchested Prinia
Prinia flavicans
685

Once obtained at Mamathe's, on 21 December 1932. This skin is with the Transvaal Museum. Also Maseru, Mountain Road (Jacot-Guillarmod 1963). Rare.

Spotted Prinia
Prinia maculosa
686

Common in Lesotho. In the lowlands associated with hillsides covered with undisturbed bush, and old dongas with some vegetation. Also common in the mountains, at least up into the mountain region.

Family   MUSCICAPIDAE: Old-world flycatchers

Seven species in Lesotho.

Spotted Flycatcher
Muscicapa striata
689

Regular and not uncommon Palaearctic visitor to the lowlands. To be seen in woodlots and in the bush along rivers. It has been seen twice in Sehlabathebe National Park, December 1977 and January 1980 (Balcomb MS).

Dusky Flycatcher
Muscicapa adusta
690

There are two sightings of this species: Qacha's Nek, 9 September 1962 (Stanford, W. *in litt.*); Maseru, on the grounds of the U.S. Embassy, 19 May 1975 (Halsted, D. *in litt.*). Rare.

Fiscal Flycatcher
Sigelus silens
698

Only a few records of this species in Lesotho: Leribe (no date); Maseru, along Caledon River, March 1963 (Jacot-Guillarmod 1963); Maseru, 17 July 1964, 28 August 1964 (Stanford, W. *in litt.*). Rare.

Cape Batis
Batis capensis
700

According to Aspinwall (1973), 'This exquisite little bird is to be seen from time to time amongst the indigenous woodland of the Maphotong Gorge. Generally found in pairs'. With no further information this species must be regarded as rare.

Chinspot Batis
Batis molitor
701

There is only one record: Mamathe's, 14 November 1947 (Jacot-Guillarmod 1963). Single record.

Fairy Flycatcher
Stenostira scita
706

Common in summer in the mountains up to 1000 metres, found breeding in December in Sanqubetu Valley (Vincent, J. *in litt.*, Symons, G. *in litt.*). Leaves the mountains in winter, and from April to September it can be seen in the lowlands, generally in areas of indigenous bush.

Paradise Flycatcher
Terpsiphone viridis
710

Seems to be a newcomer to Lesotho; so far only four sightings within a small area in Maseru: in garden of Holiday Inn, February–March 1972 (Lexander MS); at the Caledon River, below Holiday Inn, 30 November 1980 (Ryan, pers. comm.);

near the house of D. Davidson at Agricultural College, Maseru, 28 November 1982 (Davidson, D. *in litt.*); at Kent House, Pioneer Road, Maseru, 10 April 1983 (Clements, C.C. *in litt.*)

Family   MOTACILLIDAE: Wagtails, pipits, and longclaws

Nine species in Lesotho.

African Pied Wagtail
Motacilla aguimp
711

There are four sightings of this species: Mohale's Hoek (Jacot-Guillarmod 1963); Maseru, March 1972, 5 September 1974 (Lexander MS); Koma-Koma Bridge, 4 June 1979 (Bonde 1981).

Cape Wagtail
Motacilla capensis
713

Common, always near water. Resident, at least in the lowlands. Eggs collected by Jack Vincent (*in litt.*), Jarateng River, 2 December 1947; G. Symons (*in litt.*), Sanqubetu Valley, 10 December 1960.

Richard's Pipit
Anthus novaeseelandiae
716

Common in the lowland region. Specimens were collected by Jacot-Guillarmod at Mamathe's in 1932 and 1933 (Bonde 1981). In recent years it has been shown that a described race of Richard's Pipit, *A. n. editus*, from Sanqubetu Valley (Vincent 1951), belongs to another species, Mountain Pipit, *Anthus hoeschi*. Some field notes from the mountains entered as Richard's Pipits might actually belong to the Mountain Pipit. It is therefore difficult to pass any judgement on the distribution of Richard's Pipit outside the lowlands.

Mountain Pipit
Anthus hoeschi
901

In 1980, specimens in breeding condition of both this species and the former were collected at Naudesnek in the eastern Cape (Mendelsohn 1984). In Lesotho six specimens of this species were collected in Sanqubetu Valley in 1947 (Vincent 1951), from which the race *A. h. editus* was described. Also collected near Likalaneng (Quickelberge 1972). Sightings in Sehlaba-thebe National Park (Balcomb MS) of Richard's Pipit might actually refer to this species.

The Mountain Pipit is a summer visitor to be seen in the mountains of Lesotho and the north-eastern Cape above 2000 metres. Probably fairly common.

Longbilled Pipit
Anthus similis
717

Recorded commonly in the lower regions of the mountains, especially west of the Senqu. Also in the lowlands. Three specimens collected by Jacot-Guillarmod: Mamathe's, 14 July 1932; Jordane Valley, Maluti Mountains, 3 February 1933; Soosa, Maluti Mountains, 4 February 1933 (Transvaal Museum *in litt.*).

Plainbacked Pipit
Anthus leucophrys
718

One record: Agricultural Research Station, 1 December 1973 (Halsted 1974). Single record.

Rock Pipit
Anthus crenatus
721

Common in the lowlands on rocky slopes and to some extent also in the mountains up to at least 2600 metres. Eggs collected by J.A. Cottrell, Mulutsonyane Falls (misprint for Maletsunyane), 30 December 1925 (James 1970); Jacot-Guillarmod, Mamathe's, 19 January 1934 (Transvaal Museum *in litt.*).

Yellowbreasted Pipit
Anthus chloris
725

Recorded seven times from Sehlabathebe National Park: October and December 1976, January 1977, August 1978, September 1979 (Balcomb MS); January 1976, April 1982 (Coghlan, D. *in litt.*). It has been seen once in the lowlands: Koro-Koro, 31 December 1978 (Bonde 1981). It seems to be a regular visitor to the south-eastern part of Lesotho.

Orangethroated Longclaw
Macronyx capensis
727

Common in the lowlands on marshy ground, on fields next to water, and on fallow land. Seems resident. 'Not uncommon and distributed right into the central highlands' (Quickelberge 1972). Also sighted at Malefiloane Clinic (2650 metres), 3 March 1978 (Bonde 1981). Eggs collected by J. A. Cottrell, Maseru, 14 February 1924, 15 December 1925; Quthing, 10 October 1929 (James 1970).

Family LANIIDAE: Shrikes

Three species in Lesotho.

Lesser Grey Shrike
Lanius minor
731

There is one record of this species: garden in Roma, 27–29 November 1966 (MacLeay 1970). Single record.

Fiscal Shrike
Lanius collaris
732

Two subspecies can be seen in Lesotho. Time and again *L. c. subcoronatus* has been seen in the lowlands. *L. c. collaris* also is common in the lowlands. It seems to follow the rivers deep into the mountains as it has been seen at Mokhotlong, December 1947 (Vincent, J. *in litt.*); Mokhotlong, 1 December 1973; Marakabei's, 8 September 1973 (Burke MS); Linakeng River Valley (about 2250 metres), 3 August 1978 (Bonde 1981).

Redbacked Shrike
Lanius collurio
733

Uncommon Palaearctic visitor to the lowlands in summer.

Family MALACONOTIDAE: Bush shrikes

Two species in Lesotho.

Puffback
Dryoscopus cubla
740

Recorded only once: along Caledon River, below U.S. Embassy, 28 October 1973 (Halsted 1974). Single record.

Bokmakierie
Telophorus zeylonus
746

Found commonly in the lowlands, in the older gardens of Maseru and on hillsides with indigenous bush. Outside the lowlands it can be found at least up into the mountain region as long as it can find cover. Eggs collected by J.A. Cottrell, Maseru, 1 August 1919, 2 August 1920 (James 1970).

Family STURNIDAE: Starlings

Five species in Lesotho.

Pied Starling
Spreo bicolor
759

Common resident in the lowland and foothill regions. In the mountains it is common up to about 2500 metres (upper mountain valley region).

Wattled Starling
Creatophora cinerea
760

Uncommon, if not rare visitor to the lowlands usually found in flocks of *Spreo bicolor*. Seems to have decreased in number in recent years.

Plumcoloured Starling
Cinnyricinclus leucogaster
761

There is one sighting of this species: Maseru, February 1962 (Jacot-Guillarmod 1963). Single record.

Glossy Starling
Lamprotornis nitens
764

Uncommon in the lowlands. So far only seen in the Maseru-Roma area, and Leribe (Aspinwall 1973, Bonde 1981, Halsted, D. *in litt.*, Davidson, D. *in litt.*). Breeding suspected. 'Once seen collecting nesting material from my garden, Maseru, November 1970. Flew in the direction of the P.M.U. Camp' (Bass, H.G.M. *in litt.*).

Redwinged Starling
Onychognathus morio
769

Common in the lowlands, associated with the sandstone cliffs. Also common in the lower regions of the mountains (Senqu Valley and upper mountain valley regions). Recorded in Sehlabathebe National Park from December to April (Balcomb MS). Breeding recorded: Health House, Botsabela, Maseru, early November 1973; Qeme Plateau, early November 1973 (Brown MS).

Family   PROMEROPIDAE: Sugarbirds

One species in Lesotho, hypothetical.

Gurney's Sugarbird
Promerops gurneyi
774

Apart from the following statement from Jacot-Guillarmod (1963), 'So far only known from the eastern border', there is no positive evidence that this species occurs in Lesotho. Hypothetical.

Family   NECTARINIIDAE: Sunbirds

One species in Lesotho.

Malachite Sunbird
Nectarinia famosa
775

Common in the mountains, even in the high mountain region. Can also be found in the lowlands in gardens and on hillsides with indigenous bush. Breeding: Sanqubetu Valley, 10 December 1960, 13 December 1961, 14 December 1964, 11 December 1979 (Symons, G. *in litt.*).

Family   ZOSTEROPIDAE: White-eyes

One species in Lesotho.

Cape White-eye
Zosterops pallidus
796

Common in the lowlands, found in the bush along rivers and on hillsides with indigenous bush. Seems to some extent to penetrate the mountains along the rivers, though not as deeply as many other species. Reported from Makhaleng River, September 1968, (Quickelberge 1972); Ha Sempe, Quthing, 26 May 1979; along Qhoali River, 10 km from Mphaki, 7 March 1981 (Bonde 1981). Breeding: at Marakabei's, 28 October 1973 (Burke MS).

Family   PLOCEIDAE: Weavers, sparrows, bishops, widows, etc.

Twenty-two species in Lesotho.

Whitebrowed Sparrowweaver
Plocepasser mahali
799

There are two records of this species: Maseru, 2–3 July 1962. Resident. Nests: August 1962 (Stanford, W. *in litt.*); in garden in New Europa, Maseru, 7 January 1974 (Lexander MS). Rare.

House Sparrow
Passer domesticus
801

Introduced in Durban between 1893 and 1897, reached Bethlehem in 1949 and was found breeding in Ficksburg in the Orange Free State in October 1951 just across the border from Lesotho (Van der Plaat 1952). Seen at Roma and Teyateyaneng in 1954 and 1955 (Maclean 1955). Since then it has become more and more common and can be seen both in the lowlands and in the mountains (up to about 2700 metres – mountain region). New Oxbow Lodge (2650 metres), 4 March 1978 (Bonde 1981). Breeding recorded: Mokhotlong, 1 December 1973 (Burke MS).

Great Sparrow
Passer motitensis
802

David Ambrose (*in litt.*) writes, 'This species was added in the second edition (1976) of the *Guide to Lesotho*, because of a definite record from Angela Aspinwall'. Aspinwall (1973) does not mention this species. Single record, or perhaps a misidentification.

Cape Sparrow
Passer melanurus
803

Common in the lowlands and also in the mountains up to 2600 metres. Seems less dependent on human settlement than *Passer domesticus*. Breeding recorded: Mokhotlong, 1 December 1973 (Burke MS).

Greyheaded Sparrow
Passer griseus
804

Common in the lowlands throughout the year. It is the sparrow least dependent on human settlements. Seems to be spreading into the mountains along the rivers: Malefiloane Clinic, 28 February – 4 March 1978; Qhoali River, 10 km from Mphaki, 7 March 1981 (Bonde 1981).

Spectacled Weaver
Ploceus ocularis
810

David Ambrose (*in litt.*) writes, 'This species was added in the second edition (1976) of the *Guide to Lesotho*, because of a definite record from Angela Aspinwall'. Aspinwall (1973) does not mention this species, but knowing David Ambrose, I shall follow his judgement and include this species as a single record.

Spottedbacked Weaver
Ploceus cucullatus
811

Rare in the lowlands: Mohale's Hoek, Maseru, Butha-Buthe, Leribe, Teyateyaneng (Jacot-Guillarmod 1963); found along streams and occasionally on campus (Aspinwall 1973); Phatialla, 1967; Alwynskop, 1973 (Brickell 1980).

Cape Weaver
Ploceus capensis
813

Common in the mountains along the Senqu and some of its tributaries up to 3000 metres. No reports from area around Makhaleng and Senqunyane. It is uncommon in the lowlands. Breeding recorded: Lekhalabaletsi, December 1947 (Vincent, J. *in litt.*); Malibamatso and Khubelu Rivers, nesting in December (Jacot-Guillarmod 1963); golf course, Maseru, 3 December 1978; on Qhoali River, 10 km from Mphaki, 7 March 1981 (Bonde 1981).

Masked Weaver
Ploceus velatus
814

Common weaver in the lowland region, also building its nests in big trees away from water. Found in breeding dress from mid-August. Follows the rivers into the mountains and can be found up to 2600 metres.

Redbilled Quelea
Quelea quelea
821

According to Jacot-Guillarmod (1963) this species is 'generally distributed throughout the lowlands although it appears to be not quite as common as it was thirty years ago. This is at least true for the Mamathe's area'. MacLeay (1970) writes, 'Very common pest of corn crops, appears to be decreasing in numbers in Roma Valley'. Since then there have been only three sightings: Marakabei's, 1969 (Brickell 1980); P.M.U. Dam, Maseru, November 1970 (Bass, H.G.M. *in litt.*); in garden, Maseru, 6 April 1974 (Halsted, D. *in litt.*). Today this species must be regarded as rare.

Red Bishop
Euplectes orix
824

Common in the lowlands, building its nests even in small reedbeds. The males get their breeding dress from early October. Like many other species it follows the rivers deep into the mountains: Lekhalabaletsi Valley at 2900 metres, 2 December 1947 (Vincent, J. *in litt.*). Eggs collected by J.A. Cottrell, Maseru, 10 January 1922; 10 November 1922 (James 1970).

Golden Bishop
Euplectes afer
826

Common in the lowlands, but less dependent on reeds than *Euplectes orix*, staying instead in areas with fairly long grass. There is one sighting of an unusually large flock (about 1000): Koro-Koro, 18 March 1979 (Bonde 1981). Eggs collected by J.A. Cottrell, Maseru, 11 January 1922 (James 1970). Might also to some extent be spreading into the mountains along the rivers: Qhoali River, 10 km from Mphaki, 7 March 1981 (Bonde 1981).

Yellowrumped Widow
Euplectes capensis
827

Common in the mountains to 3000 metres. Is reported from Sehlabathebe National Park from July to March (Balcomb MS). Found breeding at altitudes between 2300 and 2700 metres Clancey (1957). Only one record from the lowlands: Roma, October 1965, (MacLeay 1970).

Redshouldered Widow
Euplectes axillaris
828

Only one record of this bird: Tša-Kholo, 27 March 1978, four males in the reeds (Carver 1978). Single record.

Redcollared Widow
Euplectes ardens
831

According to Jacot-Guillarmod (1963) this species should be common and widely distributed throughout the territory below 3000 metres. It seems however to have decreased in numbers in recent years. Much less common than *Euplectes progne*.

Longtailed Widow
Euplectes progne
832

This species seems to be decreasing in numbers in the lowlands, but is still common in the mountains up to 2750 metres.

Redbilled Firefinch
Lagonosticta senegala
842

This species has only been seen in Leribe: Hlotse, in garden of R. Dove, 7 September 1968, June 1970, July 1970 (Dove 1971). Rare.

Common Waxbill
Estrilda astrild
846

Common in the lowlands. There are two sightings from the mountains: 64 km by road east of Maseru on the Mountain Road about 2300 metres, 24 February to 17 March 1956 (Clancey 1957); New Oxbow Lodge, about 2650 metres, 4 March 1978 (Bonde 1981).

Swee Waxbill
Estrilda melanotis
850

The first record of this species is from Moletsane, 10 September 1968 (Quickelberge 1972). Aspinwall (1973) writes, 'Found in thick indigenous bush besides streams, as in the Maphotong Valley'. Rare.

Quail Finch
Ortygospiza atricollis
852

Common in the lowlands, mainly seen on fallow lands. Few records from the mountains: Mokhotlong District at 2600 metres, October 1946 (Vincent, J. *in litt.*); Marakabei's at 2300 metres, 16 September 1968 (Quickelberge 1972).

Orangebreasted Waxbill
Sporaeginthus subflavus
854

Rare in the lowlands, normally found near water: pond near Roma, September 1964 onwards (MacLeay 1970); Liphiring, January 1973 (Bass, H.G.M. *in litt.*); Agricultural Research Station, 27 April 1974 (Halsted, D. *in litt.*); Agricultural Research Station, 4 October 1979 (Bonde 1981).

Redheaded Finch
Amadina erythrocephala
856

Common in the lowlands, although its appearance seems somewhat erratic. Breeding recorded: Roma, 1973 (Aspinwall 1973); Mazenod, 1967 (Brickell 1980). No records from the mountains.

Family VIDUIDAE: Whydahs and widowfinches

One species in Lesotho.

Pintailed Whydah
Vidua macroura
860

Common in the lowland region, and in the mountains up to 2600 metres (Mountain Region). Seen at Sehlabathebe National Park from October to March (Balcomb MS).

Family FRINGILLIDAE: Canaries, buntings, etc.

Thirteen species in Lesotho.

Yelloweyed Canary
Serinus mozambicus
869

Only one record of this bird: Mohale's Hoek (Jacot-Guillarmod 1963). Single record.

Blackthroated Canary
Serinus atrogularis
870

Common resident in the lowlands of Lesotho. Not found in the mountains. Eggs collected by J.A. Cottrell, Maseru, 17 January 1921, 4 January 1930, 28 December 1930 (James 1970).

Cape Canary
Serinus canicollis
872

Common in the lowlands, breeding. Seems to be common in the mountains, too, wherever suitable bushy conditions occur. On Khubelu River up to 2600 metres (Jacot-Guillarmod 1963); Malefiloane Clinic, 28 February to 4 March 1978; Ha Sempe, Quthing, 26 May 1979; Koma-Koma Bridge, 4 June 1979; Qhoali River, 10 km from Mphaki, 7 March 1981 (Bonde 1981). Eggs collected by J.A. Cottrell, Maseru, 24 January 1921, 22 December 1921, 24 December 1921 (James 1970).

Drakensberg Siskin
Serinus symonsi
875

Common in the highlands of Lesotho. It ranges in the west to the foothills near Roma and in the south to Sehlabathebe National Park. Information from Thaba-Tseka area and from Mohale's Hoek and Quthing districts needed. Breeding.

Blackheaded Canary
Serinus alario
876

Apart from Jacot-Guillarmod's (1963) statement, 'Found in the northern portion of the territory', he has one record: one male about 32 km south of Mont-aux-Sources about 3300 metres. There are two other records: Liphiring, January 1973 (Bass, H.G.M. *in litt.*); Letšeng-la-Letsie, 1977 (Jilbert, J. *in litt.*). Rare.

Bully Canary
Serinus sulphuratus
877

There is only one record of this species: Kolonyama, 20 August 1973 (Halsted 1974). Single record.

Yellow Canary
Serinus flaviventris
878

Common in the mountains up to 3000 metres. Breeding. Uncommon in the lowlands and seems confined to hillsides covered with undisturbed bush, and old dongas with some vegetation.

Whitethroated Canary
Serinus albogularis
879

According to Jacot-Guillarmod (1963) it has been 'recorded once from Mamathe's, Mokhotlong, Butha-Buthe, Leribe. Probably widely distributed in the lowlands'. Only MacLeay (1970) has a sighting: Roma, November 1965. A rare species in Lesotho.

Streakyheaded Canary
Serinus gularis
881

Jacot-Guillarmod (1963) has recorded it once from Leribe, and writes further 'probably more widely distributed in the lowlands'. Aspinwall (1973) writes, 'This bird has been seen in the Maphotong Valley, but it is uncommon.' Rare.

Goldenbreasted Bunting
Emberiza flaviventris
884

Only seen a few times: Qacha's Nek; Mantšonyane (Jacot-Guillarmod 1963); Berrice Dam, Roma Valley, 18 March 1972 (Ambrose, D. *in litt.*); between Quthing and Mt. Moorosi, early November 1973 (Brown MS). Rare.

Cape Bunting
Emberiza capensis
885

Common in the lowlands on dry rocky slopes, and very common in the mountains up to 3000 metres. Eggs taken at the very source of Lekhalabaletsi River, 7 December 1947 (Vincent, J. *in litt.*).

Rock Bunting
Emberiza tahapisi
886

Common in the lowlands in the same habitat as *Emberiza capensis*, but not as numerous. There are a few records from the upper mountain valley region indicating that it might be spreading into the mountains along the rivers: Ha Sempe, Quthing, 26 May 1979; Koma-Koma Bridge, 4 June 1979 (Bonde 1981).

Larklike Bunting
Emberiza impetuani
887

Three records: Mamathe's, 30 December 1932 (Jacot-Guillarmod 1963); Koma-Koma Bridge, 3 August 1978, 4 June 1979 (Bonde 1981). A rare or overlooked species. The Koma-Koma Bridge records are interesting, as they might indicate that it follows the Senqu to the heart of Lesotho.

Index to English names

Apalis, Barthroated 645
Avocet 294

Barbet, Blackcollared 464
　Pied 465
　Redfronted Tinker 469
Bateleur 146
Batis, Cape 700
　Chinspot 701
Bee-eater, European 438
Bishop, Golden 826
　Red 824
Bittern 080
　Little 078
Bokmakierie 746
Bulbul, Redeyed 567
Bunting, Cape 885
　Goldenbreasted 884
　Larklike 887
　Rock 886
Bustard, Ludwig's 232
　Stanley's 231
Buttonquail, Kurrichane 205
Buzzard, Jackal 152
　Steppe 149

Canary, Blackheaded 876
　Blackthroated 870
　Bully 877
　Cape 872
　Streakyheaded 881
　Whitethroated 879

　Yellow 878
　Yelloweyed 869
Chat, Anteating 595
　Buffstreaked 588
　Familiar 589
　Mocking 593
　Mountain 586
　Sicklewinged 591
Cisticola, Ayres' 667
　Cloud 666
　Desert 665
　Fantailed 664
　Levaillant's 677
　Wailing 670
Coot, Redknobbed 228
Corncrake 211
Coucal, Burchell's 391
Courser, Burchell's 299
　Temminck's 300
Crake, African 212
　Spotted 214
Crane, Blue 208
　Crowned 209
　Wattled 207
Crombec, Longbilled 651
Crow, Black 547
　Pied 548
Cuckoo, Diederik 386
　European 374
　Great Spotted 380
　Jacobin 382
　Redchested 377
Curlew 289

Darter 060
Dikkop, Spotted 297
Dove, Cape Turtle 354
　Laughing 355
　Namaqua 356
　Redeyed 352
Drongo, Forktailed 541
Duck, African Black 105
　Fulvous 100
　Knobbilled 115
　Maccoa 117
　Whitebacked 101
　Whitefaced 099
　Yellowbilled 104

Eagle, Black 131
　Tawny 132
Egret, Cattle 071
　Great White 066
　Little 067
　Yellowbilled 068

Falcon, Hobby 173
　Lanner 172
　Peregrine 171
Finch, Quail 852
　Redheaded 856
Finchlark, Chestnutbacked 515
　Greybacked 516
Firefinch, Redbilled 842

Flamingo, Greater   096
   Lesser   097
Flufftail, Striped   221
Flycatcher, Dusky   690
   Fairy   706
   Fiscal   698
   Paradise   710
   Spotted   689
Francolin, Greywing   190
   Orange River   193
   Redwing   192
   Swainson's   199

Gallinule, Lesser   224
   Purple   223
Goose, Egyptian   102
   Pygmy   114
   Spurwinged   116
Goshawk, African   160
   Gabar   161
   Little Banded   159
   Pale Chanting   162
Grassbird   661
Grebe, Great Crested   006
Greenshank   270
Guineafowl, Helmeted   203
Gull, Greyheaded   315
Gymnogene   169

Hamerkop   081
Harrier, African Marsh   165
   Black   168
   Pallid   167
Heron, Blackcrowned Night   076
   Blackheaded   063
   Goliath   064
   Grey   062
   Purple   065
   Squacco   072
Honeyguide, Greater   474
   Lesser   476
   Sharpbilled   478
Hoopoe   451

Ibis, Bald   092
   Glossy   093
   Hadeda   094
   Sacred   091

Kestrel, Eastern Redfooted   180
   Greater   182
   Lesser   183
   Rock   181
Kingfisher, Giant   429
   Malachite   431
   Pied   428
Kite, Black   126
   Blackshouldered   127
   Yellowbilled   126
Korhaan, Black   239
   Blue   234
   Whitebellied   233

Lark, Clapper   495
   Longbilled   500
   Pinkbilled   508
   Redcapped   507
   Rudd's   499
   Rufousnaped   494
   Spikeheeled   506
   Thickbilled   512
Longclaw, Orangethroated   727

Martin, Banded   534
   Brownthroated   533
   House   530
   Rock   529
   Sand   532
Moorhen   226
   Lesser   227
Mousebird, Redfaced   426
   Speckled   424
   Whitebacked   425

Neddicky   681
Nightjar, European   404
   Mozambique   409

Oriole, Blackheaded   545
   European   543
Owl, Barn   392
   Cape Eagle   400
   Grass   393
   Marsh   395
   Scops   396
   Spotted Eagle   401

Pigeon, Rock   349

Pipit, Longbilled   717
   Mountain   901
   Plainbacked   718
   Richard's   716
   Rock   721
   Yellowbreasted   725
Plover, Blacksmith   258
   Crowned   255
   Grey   254
   Kittlitz's   248
   Ringed   245
   Threebanded   249
Pochard, Southern   113
Pratincole, Blackwinged   305
Prinia, Blackchested   685
   Spotted   686
   Tawnyflanked   683
Puffback   740

Quail, Common   200
   Harlequin   201
Quelea, Redbilled   821

Rail, African   210
Raven, Whitenecked   550
Reeve   284
Robin, Cape   601
   Karoo   614
Rockjumper, Orangebreasted   612
Roller, European   446
   Lilacbreasted   447
Ruff   284

Sandgrouse, Namaqua   344
Sandpiper, Common   264
   Curlew   272
   Marsh   269
   Wood   266
Secretarybird   118
Shelduck, South African   103
Shoveller, Cape   112
Shrike, Fiscal   732
   Lesser Grey   731
   Redbacked   733
Siskin, Drakensberg   875
Snake Eagle,
   Blackbreasted   143
Snipe, Ethiopian   286
   Painted   242

Sparrow, Cape   803
  Great   802
  Greyheaded   804
  House   801
Sparrowhawk, Black   158
  Little   157
  Redbreasted   155
Sparrowweaver, Whitebrowed   799
Spoonbill, African   095
Starling, Glossy   764
  Pied   759
  Plumcoloured   761
  Redwinged   769
  Wattled   760
Stilt, Blackwinged   295
Stint, Little   274
Stonechat   596
Stork, Abdim's   085
  Black   084
  White   083
  Yellowbilled   090
Sugarbird, Gurney's   774
Sunbird, Malachite   775
Swallow, European   518
  Greater Striped   526
  Lesser Striped   527
  Pearlbreasted   523
  South African Cliff   528
  Whitethroated   520

Swift, Alpine   418
  Black   412
  European   411
  Horus   416
  Little   417
  Whiterumped   415

Teal, Cape   106
  Hottentot   107
  Redbilled   108
Tern, Whiskered   338
  Whitewinged   339
Thrush, Cape Rock   581
  Kurrichane   576
  Olive   577
  Sentinel Rock   582
Tit, Ashy   552
  Southern Grey   551
Titbabbler   621,
  Layard's   622
Trogon, Narina   427
Turnstone   262

Vulture, Bearded   119
  Cape   122
  Egyptian   120
  Palmnut   147

Wagtail, African Pied   711
  Cape   713
Warbler, African Marsh   631
  Barratt's   639
  Cape Reed   635
  Willow   643
Waxbill, Common   846
  Orangebreasted   854
  Swee   850
Weaver, Cape   813
  Masked   814
  Spectacled   810
  Spottedbacked   811
Wheatear, Capped   587
White-eye, Cape   796
Whitethroat   620
Whydah, Pintailed   860
Widow, Longtailed   832
  Redcollared   831
  Redshouldered   828
  Yellowrumped   827
Woodpecker, Cardinal   486
  Ground   480
Wryneck, Redthroated   489

Index to scientific names

abdimii, Ciconia 085
abyssinica, Hirundo 527
Accipiter badius 159
   melanoleucus 158
   minullus 157
   rufiventris 155
   tachiro 160
Acrocephalus baeticatus 631
   gracilirostris 635
adsimilis, Dicrurus 541
adusta, Muscicapa 690
aegyptiacus, Alopochen 102
aethiopicus, Threskiornis 091
afer, Euplectes 826
   Parus 551
   Sphenoeacus 661
affinis, Apus 417
   Sarothrura 221
afra, Eupodotis 239
africana, Mirafra 494
africanus, Bubo 401
   Francolinus 190
   Phalacrocorax 058
aguimp, Motacilla 711
alario, Serinus 876
alba, Egretta 066
   Platalea 095
   Tyto 392
albicollis, Corvus 550
albigularis, Hirundo 520
albofasciata, Chersomanes 506
albogularis, Serinus 879
albus, Corvus 548
Alcedo cristata 431

alleni, Porphyrula 224
Alopochen aegyptiacus 102
Amadina erythrocephala 856
amurensis, Falco 180
Anas capensis 106
   erythrorhyncha 108
   hottentota 107
   smithii 112
   sparsa 105
   undulata 104
angolensis, Gypohierax 147
angulata, Gallinula 227
Anhinga melanogaster 060
Anthropoides paradisea 208
Anthus chloris 725
   crenatus 721
   hoeschi 901
   leucophrys 718
   novaeseelandiae 716
   similis 717
Apalis thoracica 645
Apaloderma narina 427
apiaster, Merops 438
apiata, Mirafra 495
Apus affinis 417
   apus 411
   barbatus 412
   caffer 415
   horus 416
   melba 418
Aquila rapax 132
   verreauxii 131
Ardea cinerea 062
   goliath 064

   melanocephala 063
   purpurea 065
ardens, Euplectes 831
Ardeola ralloides 072
Arenaria interpres 262
aridula, Cisticola 665
armatus, Vanellus 258
arquata, Numenius 289
Asio capensis 395
astrild, Estrilda 846
atricollis, Ortygospiza 852
atrogularis, Serinus 870
aurantius, Chaetops 612
auritus, Nettapus 114
avosetta, Recurvirostra 294
axillaris, Euplectes 828
ayresii, Cisticola 667
badius, Accipiter 159
baeticatus, Acrocephalus 631
Balearica regulorum 209
barbatus, Apus 412
   Gypaetus 119
barratti, Bradypterus 639
Batis capensis 700
   molitor 701
benghalensis, Rostratula 242
biarmicus, Falco 172
bicolor, Dendrocygna 100
   Spreo 759
bifasciata, Oenanthe 588
Bostrychia hagedash 094
Botaurus stellaris 080
Bradypterus barratti 639
Bubo africanus 401
   capensis 400

Bubulcus ibis   071
Burhinus capensis   297
Buteo buteo   149
   rufofuscus   152

caerulescens, Eupodotis   234
   Rallus   210
caeruleus, Elanus   127
caffer, Apus   415
caffra, Cossypha   601
cafra, Eupodotis   233
Calandrella cinerea   507
Calidris minuta   274
   ferruginea   272
calvus, Geronticus   092
cana, Tadorna   103
canicollis, Serinus   872
canorus, Cuculus   374
   Melierax   162
capensis, Anas   106
   Asio   395
   Batis   700
   Bubo   400
   Burhinus   297
   Corvus   547
   Emberiza   885
   Euplectes   827
   Macronyx   727
   Motacilla   713
   Oena   356
   Ploceus   813
   Tyto   393
capicola, Streptopelia   354
Caprimulgus europaeus   404
   fossii   409
caprius, Chrysococcyx   386
carbo, Phalacrocorax   055
carunculata, Grus   207
caudata, Coracias   447
Centropus superciliosus   391
Cercomela familiaris   589
   sinuata   591
Ceryle rudis   428
Chaetops aurantius   612
Charadrius hiaticula   245
   pecuarius   248
   tricollaris   249
Chersomanes albofasciata   506
Chlidonias hybridus   338
   leucopterus   339
chloris, Anthus   725
chloropus, Gallinula   226

Chrysococcyx caprius   386
Ciconia abdimii   085
   ciconia   083
   nigra   084
   cincta, Riparia   534
cinerascens, Parus   552
cinerea, Ardea   062
   Calandrella   507
   Creatophora   760
cinnamomeiventris, Thamnolaea   593
Cinnyricinclus leucogaster   761
Circaëtus gallicus   143
Circus macrourus   167
   maurus   168
   ranivorus   165
cirrocephalus, Larus   315
Cisticola aridula   665
   ayresii   667
   fulvicapilla   681
   juncidis   664
   lais   670
   textrix   666
   tinniens   677
Clamator glandarius   380
   jacobinus   382
Colius colius   425
   striatus   424
collaris, Lanius   732
collurio, Lanius   733
Columba guinea   349
communis, Sylvia   620
conirostris, Spizocorys   508
coprotheres, Gyps   122
Coracias caudata   447
   garrulus   446
coronatus, Vanellus   255
Corvus albicollis   550
   albus   548
   capensis   547
coryphaeus, Erythropygia   614
Cossypha caffra   601
Coturnix coturnix   200
   delegorguei   201
Creatophora cinerea   760
crenatus, Anthus   721
Crex crex   211
   egregia   212
cristata, Alcedo   431
   Fulica   228
cristatus, Podiceps   006

cubla, Dryoscopus   740
cucullata, Hirundo   526
cucullatus, Ploceus   811
Cuculus canorus   374
   solitarius   377
Cursorius rufus   299
   temminckii   300
curvirostris, Mirafra   500

delegorguei, Coturnix   201
Delichon urbica   530
Dendrocygna bicolor   100
   viduata   099
Dendropicos fuscescens   486
denhami, Neotis   231
Dicrurus adsimilis   541
dimidiata, Hirundo   523
domesticus, Passer   801
Dryoscopus cubla   740

ecaudatus, Terathopius   146
egregia, Crex   212
Egretta alba   066
   garzetta   067
   intermedia   068
Elanus caeruleus   127
Emberiza capensis   885
   flaviventris   884
   impetuani   887
   tahapisi   886
epops, Upupa   451
Eremopterix leucotis   515
   verticalis   516
erythrocephala, Amadina   856
erythrophthalma, Netta   113
Erythropygia coryphaeus   614
erythrorhyncha, Anas   108
Estrilda astrild   846
   melanotis   850
Euplectes afer   826
   ardens   831
   axillaris   828
   capensis   827
   orix   824
   progne   832
Eupodotis afra   239
   caerulescens   234
   cafra   233
europaeus, Caprimulgus   404
explorator, Monticola   582

falcinellus, Plegadis   093
Falco amurensis   180
  biarmicus   172
  naumanni   183
  peregrinus   171
  rupicoloides   182
  subbuteo   173
  tinnunculus   181
familiaris, Cercomela   589
famosa, Nectarinia   775
ferruginea, Calidris   272
flavicans, Prinia   685
flaviventris, Emberiza   884
  Serinus   878
formicivora, Myrmecocichla   595
fossii, Caprimulgus   409
Francolinus africanus   190
  levaillantii   192
  levaillantoides   193
  swainsonii   199
Fulica cristata   228
fuligula, Hirundo   529
fulvicapilla, Cisticola   681
fuscescens, Dendropicos   486

gabar, Micronisus   161
Galerida magnirostris   512
gallicus, Circaetus   143
Gallinago nigripennis   286
Gallinula angulata   227
  chloropus   226
gambensis, Pletropterus   116
garrulus, Coracias   446
garzetta, Egretta   067
Geocolaptes olivaceus   480
Geronticus calvus   092
glandarius, Clamator   380
Glareola nordmanni   305
  Tringa   266
goliath, Ardea   064
gracilirostris, Acrocephalus   635
griseus, Passer   804
Grus carunculata   207
guinea, Columba   349
gularis, Serinus   881
gurneyi, Promerops   774
Gypaetus barbatus   119
Gypohierax angolensis   147
Gyps coprotheres   122

hagedash, Bostrychia   094
hiaticula, Charadrius   245
Himantopus himantopus   295
Hirundo abyssinica   527
  albigularis   520
  cucullata   526
  dimidiata   523
  fuligula   529
  rustica   518
  spilodera   528
hoeschi, Anthus   901
horus, Apus   416
hottentota, Anas   107
hybridus, Chlidonias   338
hypoleucos, Tringa   264

ibis, Bubulcus   071
  Mycteria   090
impetuani, Emberiza   887
Indicator indicator   474
  minor   476
indicus, Urocolius   426
intermedia, Egretta   068
interpres, Arenaria   262
Ixobrychus minutus   078

jacobinus, Clamator   382
juncidis, Cisticola   664
Jynx ruficollis   489

Lagonosticta senegala   842
lais, Cisticola   670
Lamprotornis nitens   764
Lanius collaris   732
  collurio   733
  minor   731
Larus cirrocephalus   315
larvatus, Oriolus   545
layardi, Parisoma   622
leucogaster, Cinnyricinclus   761
leucomelas, Tricholaema   465
leuconotus, Thalassornis   101
leucophrys, Anthus   718
leucopterus, Chlidonias   339
leucotis, Eremopterix   515
levaillantii, Francolinus   192
levaillantoides, Francolinus   193
libonyana, Turdus   576
ludwigii, Neotis   232
Lybius torquatus   464

maccoa, Oxyura   117
Macronyx capensis   727
macroura, Vidua   860
macrourus, Circus   167
maculosa, Prinia   686
magnirostris, Galerida   512
mahali, Plocepasser   799
maurus, Circus   168
maxima, Megaceryle   429
Megaceryle maxima   429
melanocephala, Ardea   063
melanogaster, Anhinga   060
melanoleucus, Accipiter   158
melanotis, Estrilda   850
melanotos, Sarkidiornis   115
melanurus, Passer   803
melba, Apus   418
meleagris, Numida   203
Melierax canorus   162
Merops apiaster   438
Micronisus gabar   161
migrans, Milvus   126
Milvus migrans   126
minor, Indicator   476
  Lanius   731
  Phoenicopterus   097
minullus, Accipiter   157
minuta, Calidris   274
minutus, Ixobrychus   078
Mirafra africana   494
  apiata   495
  curvirostris   500
  ruddi   499
molitor, Batis   701
Monticola explorator   582
  Oenanthe   586
  rupestris   581
morio, Onychognathus   769
Motacilla aguimp   711
  capensis   713
motitensis, Passer   802
mozambicus, Serinus   869
Muscicapa adusta   690
  striata   689
Mycteria ibis   090
Myrmecocichla formicivora   595

namaqua, Pterocles   344
narina, Apaloderma   427
naumanni, Falco   183
nebularia, Tringa   270

Nectarinia famosa 775
Neophron percnopterus 120
Neotis denhami 231
  ludwigii 232
Netta erythrophthalma 113
Nettapus auritus 114
nigra, Ciconia 084
nigricans, Pycnonotus 567
nigripennis, Gallinago 286
nitens, Lamprotornis 764
nordmanni, Glareola 305
novaeseelandiae, Anthus 716
Numenius arquata 289
Numida meleagris 203
Nycticorax nycticorax 076

ocularis, Ploceus 810
Oena capensis 356
Oenanthe bifasciata 588
  monticola 586
  pileata 587
olivaceus, Geocolaptes 480
  Turdus 577
Onychognathus morio 769
Oriolus larvatus 545
  oriolus 543
orix, Euplectes 824
Ortygospiza atricollis 852
Otus senegalensis 396
Oxyura maccoa 117

pallidus, Zosterops 796
paludicola, Riparia 533
paradisea, Anthropoides 208
Parisoma layardi 622
  subcaeruleum 621
Parus afer 551
  cinerascens 552
Passer domesticus 801
  griseus 804
  melanurus 803
  motitensis 802
pecuarius, Charadrius 248
percnopterus, Neophron 120
peregrinus, Falco 171
Phalacrocorax africanus 058
  carbo 055
Philomachus pugnax 284
Phoenicopterus minor 097
  ruber 096
Phylloscopus trochilus 643

pileata, Oenanthe 587
Platalea alba 095
Plectropterus gambensis 116
Plegadis falcinellus 093
Plocepasser mahali 799
Ploceus capensis 813
  cucullatus 811
  ocularis 810
  velatus 814
Pluvialis squatarola 254
Podiceps cristatus 006
Pogoniulus pusillus 469
Polyboroides typus 169
Porphyrio porphyrio 223
Porphyrula alleni 224
Porzana porzana 214
Prinia flavicans 685
  maculosa 686
  subflava 683
Prodotiscus regulus 478
progne, Euplectes 832
Promerops gurneyi 774
Pterocles namaqua 344
pugnax, Philomachus 284
purpurea, Ardea 065
pusillus, Pogoniulus 469
Pycnonotus nigricans 567

Quelea quelea 821

ralloides, Ardeola 072
Rallus caerulescens 210
ranivorus, Circus 165
rapax, Aquila 132
Recurvirostra avosetta 294
regulorum, Balearica 209
regulus, Prodotiscus 478
Riparia cincta 534
  paludicola 533
  riparia 532
Rostratula benghalensis 242
ruber, Phoenicopterus 096
ruddi, Mirafra 499
rudis, Ceryle 428
rufescens, Sylvietta 651
ruficollis, Jynx 489
  Tachybaptus 008
rufiventris, Accipiter 155
rufofuscus, Buteo 152
rufus, Cursorius 299
rupestris, Monticola 581

rupicoloides, Falco 182
rustica, Hirundo 518

Sagittarius serpentarius 118
Sarkidiornis melanotos 115
Sarothrura affinis 221
Saxicola torquata 596
scita, Stenostira 706
Scopus umbretta 081
semitorquata, Streptopelia
  352
senegala, Lagonosticta 842
senegalensis, Otus 396
  Streptopelia 355
Serinus alario 876
  albogularis 879
  atrogularis 870
  canicollis 872
  flaviventris 878
  gularis 881
  mozambicus 869
  sulphuratus 877
  symonsi 875
serpentarius, Sagittarius 118
Sigelus silens 698
silens, Sigelus 698
similis, Anthus 717
sinuata, Cercomela 591
smithii, Anas 112
solitarius, Cuculus 377
sparsa, Anas 105
Sphenoeacus afer 661
spilodera, Hirundo 528
Spizocorys conirostris 508
Sporaeginthus subflavus 854
Spreo bicolor 759
squatarola, Pluvialis 254
stagnatilis, Tringa 269
stellaris, Botaurus 080
Stenostira scita 706
Streptopelia capicola 354
  semitorquata 352
  senegalensis 355
striata, Muscicapa 689
striatus, Colius 424
subbuteo, Falco 173
subcaeruleum, Parisoma 621
subflava, Prinia 683
subflavus, Sporaeginthus
  854
sulphuratus, Serinus 877
superciliosus, Centropus 391

swainsonii, Francolinus 199
sylvatica, Turnix 205
Sylvia communis 620
Sylvietta rufescens 651
symonsi, Serinus 875

tachiro, Accipiter 160
Tachybaptus ruficollis 008
Tadorna cana 103
tahapisi, Emberiza 886
Telophorus zeylonus 746
temminckii, Cursorius 300
Terathopius ecaudatus 146
Terpsiphone viridis 710
textrix, Cisticola 666
Thalassornis leuconotus 101
Thamnolaea cinnamomeiven-
    tris 593

thoracica, Apalis 645
Threskiornis aethiopicus 091
tinniens, Cistocola 677
tinnunculus, Falco 181
torquata, Saxicola 596
torquatus, Lybius 464
Tricholaema leucomelas 465
tricollaris, Charadrius 249
Tringa glareola 266
    hypoleucos 264
    nebularia 270
    stagnatilis 269
trochilus, Phylloscopus 643
Turdus libonyana 576
    olivaceus 577
Turnix sylvatica 205
typus, Polyboroides 169
Tyto alba 392
    capensis 393

umbretta, Scopus 081
undulata, Anas 104
Upupa epops 451
urbica, Delichon 530
Urocolius indicus 426

Vanellus armatus 258
    coronatus 255
velatus, Ploceus 814
verreauxii, Aquila 131
verticalis, Eremopterix 516
Vidua macroura 860
viduata, Dendrocygna 099
viridis, Terpsiphone 710

zeylonus, Telophorus 746
Zosterops pallidus 796